An introduction to
applied thermodynamics and energy conversion

An introduction to applied thermodynamics and energy conversion

John R. Tyldesley

Senior Lecturer in Mechanical Engineering, University of Glasgow

Longman
London and New York

To Barbara, Michael and Stephen

Longman Group Limited London

*Associated companies, branches and representatives
throughout the world*

*Published in the United States of America
by Longman Inc., New York*

© Longman Group Limited 1977

First published 1977

Library of Congress Cataloging in Publication Data

Tyldesley, John R
  An introduction to applied thermodynamics and
energy conversion.

    1. Thermodynamics.   2. Power (Mechanics)
I. Title.
TJ265.T89   621.4'021   76–56247
ISBN 0–582–44066–1
ISBN 0–582–44067–X pbk.

Printed in Great Britain by Pitman Press, Bath.

Preface

The study of energy transfer and conversion processes is not new and many texts have been written on the subject over many years. It may be asked, then, why another is regarded as necessary when the subject is already so well documented. The answer lies in the new situation presented by the world energy crisis which has resulted in a quite different economic attitude to the way in which we use our scarce energy resources and an increased awareness of their value. It has become necessary for more people to be trained in the economic use of energy, and inevitably this requires, at the very least, a rudimentary knowledge of the natural laws that govern the transfer and conversion processes. The object of this new book is to give an early undergraduate a one-year course in the most important aspects of the subject, and to do this taking full account of the new economic situation.

This book is not directed at students from any particular discipline, rather the intention is to produce a text that will prove suitable for students from a number of different fields of study. However, it is inevitable that a knowledge of mathematics of the standard of a first-year course in natural philosophy or engineering will be required, even for an introduction such as this. The book should prove suitable for both terminal and continuing courses alike, although the intention has been to produce a course that is self-contained and to which little need be added.

The text begins with an introduction to the conservation laws with which much of the work in this subject is concerned. The particular properties that are considered are mass, electric charge, momentum and energy. Each of these is examined in turn and similarities – and differences – are pointed out as appropriate. The relevant equations are developed in different forms when this will make later application easier or give improved understanding.

The second chapter is concerned with properties of materials that are required in order to apply the laws of conservation to practical problems. It is made clear in this work that many of these so-called laws are not necessarily exact and in some cases only apply to particular classes of materials. In the subsequent chapters these constitutive equations, or laws of physical properties, are used in the numerical examples that illustrate the text, and highlight the practical applications of the analyses developed in it.

The laws of thermodynamics are the main concern of the third chapter. Particular attention is given to the application of the first law to flow processes involving heat and work interactions, since these are of wide practical application. In the case of the second law the emphasis is put on the consequences arising from it, particularly in energy conversion processes. The effect of frictional dissipation is included from the outset, and throughout the development the reader is made fully aware of the economic consequences of dissipative processes. Entropy is introduced as a thermodynamic property and as a statistical parameter, the intention being to avoid divorcing macroscopic and microscopic approaches as is the case in many existing texts. An unusual, but valuable, feature of this section is the inclusion of discussions of the theory of communication and road traffic systems.

The fourth chapter is devoted to energy conversion, and the aim here is to provide the reader with a comprehensive analytical review of both well-established and new devices or machinery for energy conversion. The chapter includes sections concerned with semiconductors, fission and fusion processes, together with an introduction to the direct generation of electrical power. The chapter concludes with a brief discussion of the type of environmental considerations that may also need to be considered in practical problems.

Economics and energy conservation is the subject of the final chapter and the work here is done mainly through detailed analysis of a number of typical practical systems. The object is to provide the reader with examples of how the work of the previous chapters can be applied in conjunction with basic economics to practical systems. A study is made of integrated energy systems, the use of fuel, energy storage and the prudent use of energy resources.

This text would never have been completed had it not been for the encouragement and help given by my wife Barbara and two boys Michael and Stephen. The book is appropriately dedicated to them. The content and presentation of the material within the text owes much to the many exchanges of ideas I have had, over more than a decade, with Professor R. S. Silver of the University of Glasgow. The author is also grateful to him for the opportunities he has provided for free discussion and argument of new ideas and methods of approach to the teaching of thermodynamics and fluid mechanics.

J.R.T.
Glasgow, 1977

Contents

viii   *Contents*

Chapter 1

Conservation laws

1.1 Introduction

Thermodynamics is a study based on fundamental laws or equations which attempt to describe physical processes and interactions. In some cases the equations are known to be exact and in others may only be approximations to what is actually taking place. It is important in thermodynamics to distinguish clearly between these two types of equation or law, and in the development given here this distinction will be continually stressed.

Among the laws that may be considered to be exact is a group concerned with expressing the principle of conservation of some quantity, for example electric charge, and this chapter will be concerned primarily with them. Much of our everyday life is concerned with principles of conservation although sometimes they may not be recognised as such. As an illustration consider the following statement: 'At 09·00 hours the car park contained 30 cars. Between 09·00 and 10·00 hours 5 cars entered and 2 left. At 10·00 hours the park contained 72 cars,' – clearly nonsense, but why is this so obvious, even to a small child? Consider another example: 'A storage unit in a data link takes in $40k$ bits of information and later when instructed passes out $50k$ bits' – Again a nonsense unless the unit is malfunctioning or performing a processing function. A common feature of these two examples is that the average person with no particular expertise would expect to be able to equate the change in the content of the system to transfer across the boundary of the system and in each of the examples given this cannot be done. If the numbers are correct then it follows that the car park must be producing cars and the storage unit information – in either case not its expected function. This illustrates the essence of a conservation principle, that *changes* in the quantity of concern within a system *can* be equated exactly with transfers across the boundary of the system, the quantity itself being neither created nor destroyed.

It is interesting to form an equation expressing the principle of conservation for the car park using N for the number of cars within, N_{in} and N_{out} for the number entering and leaving. The principle may thus be written

$$\Delta N = N_{in} - N_{out} \qquad [1.1.1]$$

where ΔN indicates the change in N over the period of time considered for N_{in} and N_{out}. Correspondingly for the information storage unit, with I for the bit content of the information, we have the simpler result

$$I_{in} - I_{out} = 0 \qquad [1.1.2]$$

for the period of time considered.

Consider now a more complex storage unit which can take in information on a rapid, continuous basis, stores selected information for varying periods of time, and passes out information rapidly. Its operation over some period of time Δt can now be represented by analogy with the car park by the equation

$$\Delta I = I_{in} - I_{out} \qquad [1.1.3]$$

or dividing by Δt

$$\frac{\Delta I}{\Delta t} = \frac{\Delta I_{in}}{\Delta t} - \frac{\Delta I_{out}}{\Delta t} \qquad [1.1.4]$$

If Δt is small and the process of bit transfer sufficiently rapid it may be better to write this as a differential equation in the form

$$\frac{dI}{dt} = \frac{dI_{in}}{dt} - \frac{dI_{out}}{dt} \qquad [1.1.5]$$

The equations describing the operation of even more sophisticated storage systems with multi-entry and dispatch facilities can be built up from this equation and, similarly, equations for the movement of vehicles in a busy town with many access roads can be readily devised.

However, these two examples have a common feature that makes development of a conservation equation apparently easier than in many other cases, and that is that the quantity considered, i.e. cars or bits, exists in unit lumps. In other cases, for example chemical pollutant in a river, the quantity may be continuously distributed and can only be described in terms of a concentration or density type of function. Despite this difference the development of corresponding conservation equations is in fact no more difficult, as will be shown in later sections of this chapter.

1.2 Mass

The object of this and subsequent sections is to take a particular conservation law and develop a simple framework to enable the law to be expressed in a mathematical form. This is necessary so that when physical processes are analysed the results are known to be compatible with the physical laws under which real processes take

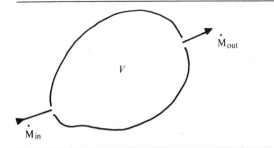

Fig. 1.2.1. Mass transfer system.

place. If the results of any analysis are at variance with the conservation laws then the process cannot exist in the real world and the analysis has not served any useful purpose.

In its simplest terms the principle of conservation of mass implies two things. The first is that mass can neither be created nor destroyed, and the second is that the mass contained within a fixed region of space can only be changed by a flux of mass through the boundary of the region. The principle is *not* valid in situations where relativistic phenomena are significant, but most engineering situations do not come into this category and there the principle may be regarded as generally valid.

Consider now the system shown in Fig. 1.2.1, where V is a fixed volume of space having mass M and where there is an input of mass at the rate $\dot{M}_{in}$ and an outlet flux $\dot{M}_{out}$. In words the principle can be expressed as

mass input rate − mass output rate = mass storage rate

and in terms of the symbols M, $\dot{M}_{in}$ and $\dot{M}_{out}$ as

$$\frac{dM}{dt} = \dot{M}_{in} - \dot{M}_{out} \qquad [1.2.1]$$

In the form of equation [1.2.1] the law is still not in the best form for application except in the simplest of practical problems where $\dot{M}_{in}$ and $\dot{M}_{out}$ are known directly. In most cases the mass fluxes must be first found from knowledge of the flow velocities, flow areas and the properties of the material crossing the boundary. Consider then a small section of the boundary shown in Fig. 1.2.2 where mass is passing through the boundary. The velocity of the material is taken to be v normal to the surface, its density ρ and the process is examined over a small time element δt. In this time interval the material originally on the boundary surface moves a distance δl and inspection shows that the mass which has crossed the boundary is given by

$$\delta M = \rho \, \delta A \, \delta l \qquad [1.2.2]$$

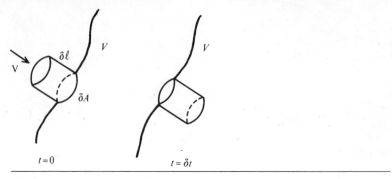

Fig. 1.2.2. Elemental mass transfer.

The rate at which mass crosses the boundary, $\delta \dot{M}$, is the limit $(\delta M / \delta t)_{\delta t \to 0}$, i.e.

$$\delta \dot{M} = \rho \, \delta A \mathrm{v} \qquad [1.2.3]$$

For a large area A the net mass flux $\dot{M}$ is obtained by integration over the surface area, i.e.

$$\dot{M} = \int_A \rho \mathrm{v} \, \mathrm{d}A \qquad [1.2.4]$$

In many cases the velocity and density can be taken to be constant over the area and the expression for $\dot{M}$ then becomes

$$\dot{M} = \rho A \mathrm{v} \qquad [1.2.5]$$

Equation [1.2.5] turns out to be a most useful result for many practical applications.

Example 1.2.1
Water flows through a tube of varying cross-section and the velocity may be considered to be sensibly uniform over any particular cross-section. The mass flow rate is 9·5 kg/sec. Find the velocity of flow at two cross-sections having diameters of 100 mm and 30 mm.

Data

$\dot{M} = 9 \cdot 5$ kg/sec

$d_1 = 100 \, \text{mm}, \qquad d_2 = 30 \, \text{mm}$

$\rho \ = 1000 \ \text{kg/m}^3$

Analysis

$$\dot{M} = \rho A v = \rho\left(\frac{\pi d^2}{4}\right)v$$

Therefore

$$v = \frac{4\dot{M}}{\pi\rho d^2}$$

Calculation

$$v_1 = \frac{4 \times 9 \cdot 5}{\pi \times 1000 \times (0 \cdot 1)^2} = \mathbf{1 \cdot 21\ m/sec}$$

$$v_2 = \frac{4 \times 9 \cdot 5}{\pi \times 1000 \times (0 \cdot 03)^2} = \mathbf{13 \cdot 4\ m/sec}$$

Equations [1.2.1] and [1.2.4] can now be used together to give a useful equation expressing the principle of conservation of mass for a system having a mass M and bounded by a surface of area A. Combining these two equations we obtain

$$\frac{dM}{dt} = \int_A \rho v\ dA \qquad\qquad [1.2.6]$$

Note that in expressing the mass inflow and outflow rates it is not necessary to distinguish between inlet and outlet areas in the integral term since the integral is necessarily positive for the former and negative, as required, for the latter.

1.3 Charge

The development of equations expressing the conservative nature of electric charge can be formulated almost by analogy with the developments in Section 1.2. This is possible because, like mass, charge cannot be created or destroyed and the amount of charge in a fixed region of space can only be changed by a flux of mass across the boundary of that region. In addition, electric charge is only associated with lumps of matter and has no value in a region devoid of matter. If a charged particle or lump of matter moves with velocity v, then the charge moves with the particle of matter and thus the electric charge may be loosely thought of as having the velocity v, although strictly the velocity is that of the material upon which the charge resides. However, an important difference is that the charge on a particle can be positive, negative or zero. By convention the electron charge is taken to be negative and thus an atom which has lost an electron is positively charged. A flux of positively charged particles constitutes an electric current which by convention is said to be in the same direction as the charge flux. In a metallic conductor the current flow is the result of a flux of electrons, negatively charged, and consequently the electron flux is in the opposite direction to that of the current as usually defined. In some materials, for example an electrolyte, the current is the result of two fluxes: one of positively charged ions moving in the same direction as the current, the other of negatively charged ions moving in the opposite direction. Each contributes equally to the total flux or electric current.

When dealing with electric charge the mass density ρ is replaced by the charge density ρ_c which is simply the charge per unit volume and may, like the particle charge, be positive, negative or zero. A zero value for ρ_c does not necessarily indicate the absence of charged particles. When current flows in a metallic conductor the net charge in any small volume is zero but, providing the volume is large compared to molecular dimensions, it contains many moving electrons and stationary atoms with negative and positive charges respectively. In order to avoid ambiguity from this, it is convenient now to define two charge densities ρ_c^+ and ρ_c^- for the positively and negatively charged particles, with ρ_c^- being by definition a zero or negative number, thus

$$\rho_c = \rho_c^+ + \rho_c^- \qquad\qquad [1.3.1]$$

When charged particles cross a surface there are two contributions to the net flux, the one from the passage of positively charged particles and the other from the passage of the negatively charged particles. Using I^+ and I^- for these fluxes we have, by analogy with the development of equation [1.2.4],

$$I^+ = \int_A \rho_c^+ v^+ \, dA$$

$$[1.3.2]$$

$$I^- = \int_A \rho_c^- v^- \, dA$$

where v^+ and v^- are the average velocities of the positively and negatively charged particles respectively. The net current I is the algebraic sum of I^+ and I^-, giving

$$I = \int\limits_A (\rho^+ v^+ + \rho^- v^-)\, dA \qquad\qquad [1.3.3]$$

In many practical cases v^+ and v^- may like ρ^+ and ρ^- be taken to be constant over the area A and for these the equation becomes

$$I = (\rho^+ v^+ + \rho^- v^-)A \qquad\qquad [1.3.4]$$

Example 1.3.1
A sample of a copper conductor contains $8 \cdot 5 \times 10^{28}$ free electrons per cubic metre each having an electric charge of $-1 \cdot 6 . 10^{-19}$C (Coulomb). Find the average or drift velocity of the electrons through a copper wire 1 mm in diameter carrying a current of 1 A (1 C/sec).

Data

$n = 8 \cdot 5 \times 10^{28}/\text{m}^3$

$q = -1 \cdot 6 \times 10^{-19}\,\text{C}$

$d = 1 \times 10^{-3}\,\text{m}$

$I = 1\,\text{A}$

Analysis

$$I = (\rho_c^+ v^+ + \rho_c^- v^-)A, \qquad v^+ = 0$$

$$I = \rho_c^- v^- A = (nq) \times \left(\frac{\pi\, d^2}{4}\right) v^-$$

Therefore

$$v^- = \frac{4I}{\pi n q\, d^2}$$

Calculation

$$v^- = -\frac{4 \times 1}{\pi \times 8 \cdot 5 \times 10^{28} \times 1 \cdot 6 \times 10^{-19} \times 10^{-6}}$$

$$= -\mathbf{93 \cdot 6 \times 10^{-6}\,m/sec}$$

Example 1.3.1 illustrates that the electrons move through the conductor in the opposite sense to that of the electric current direction and with an exceedingly low drift velocity.

The conservation of electric charge may be expressed in an equation that may be deduced by analogy with equation [1.2.6] in Section 1.2. A slight difference arises because electric charge can be carried by both positive and negative ions in an electrolyte. The equation thus contains two integral terms on the right-hand side and is

$$\frac{dQ}{dt} = \int_A (\rho^+ v^+ + \rho^- v^-) dA \qquad [1.3.5]$$

In this equation dQ/dt is the rate at which the charge contained within the volume increases, and v^+ and v^- are the inward velocity of the positive and negative charges respectively.

1.4 Momentum

The momentum of mass M moving with velocity v is defined to be the product $P = Mv$. It is therefore a vector quantity and in this sense is rather different from mass and charge considered in sections 1.2 and 1.3. When dealing with the conservation of momentum each of the components has to be considered separately and the conservation principle applied to each component in turn. In many practical problems it is convenient to use rectangular Cartesian coordinates when dealing with momentum, and in this system the components of momentum of the mass M would be

$$P_x = Mv_x, \qquad P_y = Mv_y, \qquad P_z = Mv_z$$

If there are a number of masses within the system then the total momentum of the system P is defined by it having components given by

$$P_x = \sum Mv_x, \qquad P_y = \sum Mv_y, \qquad P_z = \sum Mv_z$$

where each summation is taken over all the individual masses within the system. For such a system which is completely isolated and can have no interaction with any other system outside, the principle of conservation of momentum states that each of the components P_x, P_y, P_z remains constant, i.e.

$$P_x = \text{constant}, \qquad P_y = \text{constant}, \qquad P_z = \text{constant}$$

The principle does not in any way prohibit interchange of momentum between the masses within the system, but does ensure that in such interactions the total momentum remains constant.

Such a statement of the law is of little use unless the required isolated system can be identified without ambiguity. In the case of mass and charge this was comparatively simple. In the case of mass conservation the requirement was that no mass crossed the boundary of the system. For an isolated system of charges it was necessary that no charged particles crossed the boundary, and in addition there had to be no electromagnetic radiation across the boundary of the system. The situation with momentum is very similar to that encountered when dealing with the conservation of electric charge. Two quite different processes can be identified which can result in a change of momentum of a mass or system of masses.

An example will quickly identify the two processes. Consider the system consisting of an iron pendulum which is initially stationary and therefore has zero momentum. A lump of clay is hurled at it, sticks to it, and the result is that the pendulum system, which now includes the clay, has a non-zero momentum. The conclusion to be drawn is that momentum can be transferred to a system by a transfer process occurring as a result of a mass transfer across a boundary surrounding the system. Alternative methods for increasing the momentum of the pendulum would be to tap it with a hammer or to bring a powerful magnet close to it. The interaction in each of these cases would normally be described in terms of a force acting between the finger and the pendulum in the former, and in the latter, the interaction would be related to a magnetic attractive or repulsive force between the magnet and the iron pendulum. The distinction, then, is between two processes: one in which the momentum transfer is associated with a mass transfer, and the other in which the transfer is quite independent of any mass transfer. An isolated system in this case must therefore be one in which no mass crosses the boundaries, and it is also necessary that there is no force interaction across the boundary surface.

For a system which is not isolated the change in momentum of the system will be the sum of the momentum changes resulting from each type of process. Each of these processes is conventionally dealt with a little differently as will now be shown.

Consider a region of space of mass M, volume V and having momentum $\mathbf{P}$. If a force $\mathbf{F}$ acts on the mass M then from Newton's law the rate of change of momentum $d\mathbf{P}/dt$ is proportional to $\mathbf{F}$ and in the normal system of international units the constant is arranged to be unity. Thus

$$\mathbf{F} = \frac{d\mathbf{P}}{dt} \qquad\qquad [1.4.1]$$

An alternative way of changing the momentum within V is to transfer a mass, say δM, having a momentum $\delta \mathbf{P}$ into the region. The principle of conservation of momentum then gives the new momentum of the region as $(\mathbf{P} + \delta \mathbf{P})$, given by

$$\mathbf{P} + \delta \mathbf{P} = \mathbf{P} + \delta M \mathbf{v}$$

The corresponding *rate* of change of momentum that occurs if mass is introduced into V on a continuous basis is $d\mathbf{P}/dt$ given by

$$\frac{d\mathbf{P}}{dt} = \dot{M}\mathbf{v} \qquad [1.4.2]$$

where $\dot{M}$ is the mass transfer rate.

If now the region is acted on by the force $\mathbf{F}$ and at the same time a flow into the region is taking place, the *total* rate of change of momentum $d\mathbf{P}/dt$ is the algebraic sum of the rates in equations [1.4.1] and [1.4.2], i.e.

$$\frac{d\mathbf{P}}{dt} = \mathbf{F} + \dot{M}\mathbf{v} \qquad [1.4.3]$$

The momentum equation [1.4.3] can be applied directly to a system having a number of inlets and outlets, over each of which the velocity may be assumed to be uniform. However, care is necessary, to distinguish clearly between inlet sections and outlet sections in order to achieve the correct algebraic sum for the momentum transfer associated with mass transfer. Thus, for a system with inlet and outlet mass transfers equation [1.4.3] when applied becomes

$$\frac{d\mathbf{P}}{dt} = \mathbf{F} + \sum_{in} \dot{M}\mathbf{v} - \sum_{out} \dot{M}\mathbf{v} \qquad [1.4.4]$$

where $\mathbf{P}$ is the momentum of the whole system within the volume V, $\mathbf{F}$ is the algebraic sum of the external forces acting on the mass within V, and $\sum_{in}$, $\sum_{out}$ are summations over the inlets and outlets respectively. It will prove convenient to differentiate between two types of forces that contribute to the net force $\mathbf{F}$ acting on the mass. We distinguish *surface forces* associated with pressure, friction and surface tension for example, from volume or *body forces* which act directly on the elements of mass within V. Typical examples here are gravitational and electromagnetic forces. The two types of force will be denoted by $\mathbf{F}_S$ and $\mathbf{F}_V$ respectively. The force $\mathbf{F}$ in equation [1.4.4] is the algebraic sum of $\mathbf{F}_S$ and $\mathbf{F}_V$ and when this is introduced into equation [1.4.4] the final result is

$$\frac{d\mathbf{P}}{dt} = \mathbf{F}_S + \mathbf{F}_V + \sum_{in} \dot{M}\mathbf{v} - \sum_{out} \dot{M}\mathbf{v} \qquad [1.4.5]$$

If the region of space being considered does not have distinct inlets and outlets, or if the velocity across their cross-section is not uniform, it is often more convenient to use an integral form of equation [1.4.4]. However, it must be stressed that in making such alternative formulations no new information is being added and the resulting equation will be simply an alternative presentation of the same basic principle.

The analysis is begun by considering a small element of surface δA across which there is a mass flux with velocity **v**. The element of momentum flux resulting from this, $\delta\dot{\mathbf{P}}$, is given by

$$\delta\dot{\mathbf{P}} = \delta\dot{M}\mathbf{v} = \rho(\mathbf{v}.\delta\mathbf{A})\mathbf{v} \qquad [1.4.6]$$

where $\delta\mathbf{A}$ is a vector having the magnitude of δA and a direction normal to δA and into the surface. Examination of equation [1.4.6] shows that the sign of the right-hand term is positive for flows into V and negative for flows out of V, and this integration over the whole surface of V will itself take account of inlet and outlet momentum transfers without any further requirement to distinguish them. The total momentum flux into V, **P,** may thus be obtained by integration of equation [1.4.6] over the surface of V and is

$$\dot{\mathbf{P}} = \int_V d\dot{\mathbf{P}} = \int_A \rho(\mathbf{v}.d\mathbf{A})\mathbf{v} \qquad [1.4.7]$$

$\dot{\mathbf{P}}$ is the corresponding integral form of the combined second and third terms on the right of equation [1.4.4] and it now remains to look into the left-hand term representing the rate of change of momentum of the mass M contained within V. The total momentum contained within V is **P**, given by

$$\mathbf{P} = \int_V \rho\mathbf{v}\,dV$$

and it follows that the rate of change of **P** is

$$\frac{d\mathbf{P}}{dt} = \frac{d}{dt}\left(\int_V \rho\mathbf{v}\,dV\right) \qquad [1.4.8]$$

Normally the mass within V may be taken to be continuous and corresponding properties hold for ρ and **v**. In such conditions it is then permissible to interchange the order of differentiation and integration in equation [1.4.8] if this is desirable. Thus, an alternative presentation is

$$\frac{d\mathbf{P}}{dt} = \int_V \left(\frac{d}{dt}(\rho\mathbf{v})\right)dV \qquad [1.4.9]$$

With the form of the momentum flux and change terms now determined, equation [1.4.4] can now be written in the integral form as

$$\frac{d}{dt}\left(\int_V \rho\mathbf{v}\,dV\right) = \mathbf{F}_S + \mathbf{F}_V + \int_A \rho(\mathbf{v}.d\mathbf{A})\mathbf{v} \qquad [1.4.10]$$

In this equation distinction has again been made between surface and volume forces.

For many practical problems, particularly those concerned with the thermodynamic aspects of steady fluid flow, there is a third form of presentation which proves to be more convenient in practice. The derivation of this is begun by considering the application of the momentum conservation principle using equation [1.4.4] and applying it to a small elemental stream tube as shown in Fig. 1.4.1.

A stream tube is simply a volume element whose surface, apart from the ends, is bounded by streamlines of the fluid flow. As shown in the diagram, the end surfaces, distance δl apart, have areas $(A, A + \delta A)$. The corresponding pressures are $(p, p + \delta p)$ and the velocities $(v, v + \delta v)$. The axial component of the surface force F_s acting on the outer surface independently of the pressure is f_s/unit mass, and the axial component of the volume force is f_v/unit mass. Applying equation [1.4.4] to this system we get to the first order in small quantities.

$$\frac{d}{dt}(\rho A\,\delta l v) = pA - (p + \delta p)(A + \delta A) - p\,\delta A -$$
$$- \rho A\,\delta l(f_s + f_v) + (\rho A v)v - (\rho A v)(v + \delta v)$$
$$= -A\delta p - \rho A\,\delta l(f_s + f_v) - \rho A v\,\delta v$$

or

$$\frac{1}{\rho A}\frac{d}{dt}(\rho A v)\delta l = -\frac{\delta p}{\rho} - f_s\,\delta l - f_v\,\delta l - v\,\delta v \qquad [1.4.11]$$

For steady flows where the left-hand term is zero this may be written

$$v\,\delta v + \frac{\delta p}{\rho} + f_s\,\delta l + f_v\,\delta l = 0$$

or

$$\delta\!\left(\frac{v^2}{2}\right) + \frac{\delta p}{\rho} + f_s\,\delta l + f_v\,\delta l = 0 \qquad [1.4.12]$$

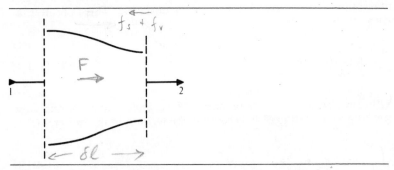

Fig. 1.4.1. The elemental stream tube.

It is interesting at this point to look in a little more detail at the two terms $f_S \, \delta l$ and $f_V \, \delta l$ before proceeding further. The first of these represents the work done by the surface force f_S on unit mass of fluid when it is moved a distance δl along the stream tube. In most fluid flows the force f_S will be the result of frictional or viscous forces acting on the surface of the stream tube and it is tempting at this point to regard the product $f_S \, \delta l$ as a term representing frictional dissipation alone. However, consider the example of a car accelerating slowly on an icy road with the drive wheels slipping on the surface. In this case if the car is accelerating it is clear that a part of the work being done by the frictional forces acting on the tyres is useful and is producing acceleration of the vehicle. Some of the work being done by the friction force, however, is purely dissipative and results in an increased tyre temperature only. The situation is similar in the case of the fluid flow and the work term $f_S \, \delta l$. Part may be dissipative work and the remainder useful work done on or by the fluid element being considered. The subdivision can only be made from knowledge of the complete flow system, and in the general case nothing more can be said. However, it is useful to represent the two distinctive processes by two terms $\Delta \Phi$ and ΔW to indicate the distinction between a dissipative and a useful work effect. Thus we write

$$f_S \, \delta l = \Delta \Phi + \Delta W \qquad [1.4.13]$$

where $\Delta \Phi$ represents an element of dissipated work and ΔW an element of useful work. The notation $\Delta \Phi$ and ΔW is adopted to indicate that neither $\Delta \Phi$ nor ΔW is an exact differential.

The second of these terms $f_V \, \delta l$ is the work done on unit mass of fluid moving a distance δl along the stream tube by the forces acting on all particles of fluid throughout the volume of the stream-tube element. In most flows the force f_V will be the result of a potential force field such as a gravitational or electric field. We shall restrict the subsequent discussion to such cases where there exists a potential ϕ for the force field f_V. The potential ϕ is usually defined such that the work done in moving a mass M in a gravitational field from a point at potential ϕ to another point where the potential is $(\phi + \delta \phi)$ is $M \, \delta \phi$, and thus the work done per unit mass is just $\delta \phi$. Consequently, if f_V is the result of a potential ϕ then

$$f_V \, \delta l = \delta \phi \qquad [1.4.14]$$

Using equations [1.4.13] and [1.4.14] we may now rewrite the original equation [1.4.12] in a form that will be found to be extremely useful later.

$$\delta \left(\frac{v^2}{2} \right) + \frac{\delta p}{\rho} + \Delta \Phi + \Delta W + \delta \phi = 0 \qquad [1.4.15]$$

In this equation $\delta(v^2/2)$ is the change in kinetic energy per unit mass, δp the change in pressure, $\Delta\Phi$ the frictional dissipation, ΔW the useful work output and $\delta\phi$ the change in potential – all between two sections of the stream tube a distance δl apart in the flow direction. Integration along the stream tube between two cross-sections 1 and 2 gives finally

$$\int_1^2 d\left(\frac{v^2}{2}\right) + \int_1^2 \frac{dp}{\rho} + \int_1^2 \Delta\Phi + \int_1^2 \Delta W + \int_1^2 d\phi = 0$$

or

$$\frac{(v_2^2 - v_1^2)}{2} + \int_1^2 \frac{dp}{\rho} + \Phi_{12} + W_{12} + (\phi_2 - \phi_1) = 0 \qquad [1.4.16]$$

A point to note here is that Φ_{12} is the net frictional dissipation which occurs when a unit mass of fluid passes between sections 1 and 2. The term W_{12} is the corresponding net useful work input, or output if negative, that takes place. In a fluid flow in which the density varies along the stream tube it is not possible to integrate the term $\int_1^2 (dp/\rho)$. Fortunately, for many flows commonly encountered such as flow of a liquid or even a gas at low velocity, the variation of density may well be insignificant and this term can then be integrated directly to give $(p_2 - p_1)/\rho$. Substitution in equation [1.4.16] yields finally for such type of flow

$$\frac{(v_2^2 - v_1^2)}{2} + \frac{(p_2 - p_1)}{\rho} + \Phi_{12} + W_{12} + (\phi_2 - \phi_1) = 0 \qquad [1.4.17]$$

The starting point for much of the later developments and practical examples will be one of either equations [1.4.16] or [1.4.17], the former in the general case and the latter in the rather specialised situation where variations in density may be neglected. It will also prove possible in many flow geometries to neglect the effect of any variations in potential ϕ, again producing considerable simplification. Except in the case of work input or output devices, such as compressors and turbines, the work term W_{12} will be zero. Unfortunately, the frictional dissipation term Φ_{12} presents no such simplification. No practical engineering flow can exist without dissipation and hence it follows that this term can never strictly be ignored. In practice it proves extremely difficult to make an exact allowance for frictional effects, and this has led to the development of a number of specialised techniques for dealing with it. These techniques include, surprisingly often, the most easy solution of all, which is to ignore it. Sometimes the effect of the term Φ_{12} is small and in these cases it is often in fluid geometries where there is no work effect when the equation reduces to that deduced by Bernoulli and widely called after him. The Bernoulli equation is therefore

$$\frac{(v_2^2 - v_1^2)}{2} + \frac{(p_2 - p_1)}{\rho} + (\phi_2 - \phi_1) = 0 \qquad [1.4.18]$$

Its use *must* be restricted to those flows under which the conditions outlined above hold. These conditions are:

(a) No work input or output effect.
(b) Density variation can be neglected.
(c) Frictional dissipation effects negligible.

Unless all these conditions are satisfied the equation must *not* be used and instead must be replaced by one or more of the more general equations given previously.

Example 1.4.1
Water flows steadily through a horizontal pipe of 0·03 m diameter at a velocity of 6·5 m/sec. At some point the pipe diameter reduces to 0·02 m and tests show that the pressures before and after the reduction in area are 250 kN/m² and 150 kN/m². Find the frictional dissipation, and the pressure that would have been found in the reduced diameter section if the flow had been frictionless.

Data

$d_1 = 0 \cdot 03$ m, $\qquad d_2 = 0 \cdot 02$ m
$p_1 = 250$ kN/m², $\qquad p_2 = 150$ kN/m²
$v_1 = 6 \cdot 5$ m/sec
$\rho = 1000$ kg/sec

Analysis

$$\dot{M} = \rho A_1 v_1 = \rho A_2 v_2$$

Therefore

$$v_2 = \frac{A_1}{A_2} v_1 = \left(\frac{d_1}{d_2}\right)^2 v_1$$

From equation [1.4.17] with $W_{12} = 0$ and $\phi_1 = \phi_2$ we have

$$\Phi_{12} = \frac{v_1^2 - v_2^2}{2} + \frac{(p_1 - p_2)}{\rho}$$

and thus Φ_{12} can be found.

If the flow had been frictionless $\Phi_{12} = 0$ and thus

$$p_2 = p_1 + \frac{\rho(v_1^2 - v_2^2)}{2}$$

Calculation

$$v_2 = \left(\frac{0 \cdot 03}{0 \cdot 02}\right)^2 \times 6 \cdot 5$$

$$= 14 \cdot 63 \text{ m/sec}$$

$$\Phi_{12} = \frac{(6 \cdot 5^2 - 14 \cdot 63^2)}{2} \times 10^{-3} + \frac{(250 - 150)^*}{1000}$$

$$= \mathbf{0 \cdot 0141 \text{ kJ/kg}}$$

* *Note:* The factor 10^{-3} is included in this term so that both terms are in kJ/kg.

For frictionless flow

$$p_2 = 250 + \frac{1000}{2}(6 \cdot 5^2 - 14 \cdot 63^2) \times 10^{-3}$$

$$= \mathbf{164 \cdot 1 \text{ kN/m}^2}$$

Example 1.4.2
A liquid of density 750 kg/m³ is to be pumped from a ground level reservoir into an open storage tank at a height of 30 m. The pipe connecting the pump and tank is 0·01 m in diameter. If the mass flow required is 1 kg/sec, find the minimum pressure at the pump exit and the minimum power input to the pump. Assume atmospheric pressure to be 101 kN/m².

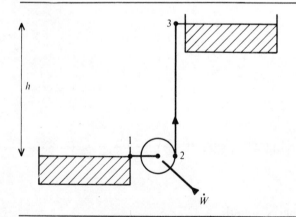

Data

$p_1 = p_3 = 101$ kN/m^2

$v_1 = 0$

$\rho = 750$ kg/m^3

$d = 0 \cdot 01$ m

$h = 30$ m

$\dot{M} = 1$ kg/sec

Analysis

$$\dot{M} = \rho A v$$

Therefore

$$v_2 = v_3 = \frac{\dot{M}}{\rho A}$$

From equation [1.4.17] for the flow between 2 and 3 we have

$$\frac{(v_3^2 - v_2^2)}{2} + \frac{(p_3 - p_2)}{\rho} + \Phi_{23} + (\phi_2 - \phi_1) = 0$$

however $v_3 = v_2$ and $(\phi_2 - \phi_1) =$ difference in gravitational potential between 2 and 3 which is *gh*.

Thus we have

$$p_2 = p_3 + \rho gh + \Phi_{23}$$

and the minimum is when $\Phi_{23} = 0$, i.e. frictional effects are negligible. Therefore

$$p_{2_{\min}} = p_3 + \rho gh$$

For the process from 1 to 2 we have from equation [1.4.17]

$$W_{12} = \frac{(v_1^2 - v_2^2)}{2} + \frac{(p_1 - p_2)}{\rho} - \Phi_{12}$$

with $v_1 = 0$ and $v_2 = \dot{M}/\rho A$.

The pump requires a work input and hence the work *input* per unit mass $W = -W_{12}$. Therefore

$$W = \frac{1}{2} \left(\frac{\dot{M}}{\rho A} \right)^2 + \frac{(p_2 - p_1)}{\rho} + \Phi_{12}$$

This is a minimum when $\Phi_{12} = 0$ and $p_2 = p_{2_{\min}}$. Hence

$$\dot{W}_{\min} = \frac{1}{2} \left(\frac{\dot{M}}{\rho A} \right)^2 + \frac{(p_{2_{\min}} - p_1)}{\rho}$$

Minimum power input $P_{min} = \dot{M}W_{min}$. Therefore

$$P_{min} = \dot{M}\left\{\frac{1}{2}\left(\frac{\dot{M}}{\rho A}\right)^2 + \frac{(p_{2_{min}} - p_1)}{\rho}\right\}$$

Calculation

$$p_{2_{min}} = 101 + 750 \times 9\cdot81 \times 30 \times 10^{-3}$$
$$= 321\cdot7 \text{ kN/m}^2$$

$$P_{min} = 1 \times \left\{\frac{1}{2}\left(\frac{1 \times 4}{750 \times \pi \times 0\cdot01^2}\right)^2 \times 10^{-3} + \frac{(321\cdot7 - 101)}{750}\right\}$$
$$= 0\cdot438 \text{ kJ/sec}$$
$$= \mathbf{0\cdot438 \text{ kW}}$$

1.5 Energy

Little explanation is necessary for a complete understanding of what is meant by the mass, charge or momentum content associated with a body or a system. Energy, however, is rather different mainly because it can exist in different forms, some of which are by no means immediately obvious. Perhaps the most easily recognised and widely understood form of energy is that associated with the motion of a body, namely its *kinetic energy*. For a mass M moving with velocity v the kinetic energy (k.e.) is defined by

$$(\text{k.e.}) = \tfrac{1}{2}Mv^2$$

Earlier work in mechanics will also have introduced the idea of potential energy due to gravitational attraction and also the notion that potential energy can be converted to kinetic energy, as for example when a mass falls from a high point above the earth to a lower one. When the mass falls its gravitational potential energy decreases and its accelerating motion results in a corresponding increase in its kinetic energy. In the absence of a drag force due to friction, the falling process results in no overall increase or decrease in the sum total of kinetic and potential energy – clearly a manifestation of some principle of energy conservation. However, consider now the next stage in the process. The mass hits the earth and thereafter its kinetic energy remains zero. Immediately before the collision the kinetic energy had a value greater than zero and a casual observer might deduce that the process of collision had destroyed energy. A more astute observer might notice that one effect of the collision was that the mass and the earth in the neighbourhood were now hotter than before the collision, and he might then postulate that this could be an indication that an energy change had occurred within the materials themselves as a result of some energy-transfer process. With this idea of an *internal energy* of the materials the transfer processes associated with the fall and collision could then be written

$$\text{potential energy} \xrightarrow{\text{fall}} \text{kinetic energy} \xrightarrow{\text{collision}} \text{internal energy}$$

This flow diagram shows two energy-transfer processes and the directions in which they have taken place for this particular system. Clearly, the first of these is reversible and if the mass is thrown upwards its kinetic energy reduces and potential energy increases. This raises the obvious query. Can an internal energy be used to increase the kinetic energy of a mass? A number of simple experiments can be devised to demonstrate that this is in fact the case. For example a hot stone placed in a bath of cold water will cause convective motion of the water, i.e. an increase in kinetic energy, and this will not happen if the water and stone are at the same degree of hotness or coldness. A balloon filled with hot air rises in the atmosphere, producing an increase in both kinetic *and* potential energy. Nothing happens if the balloon is filled with air at atmospheric temperature. Many demonstrations of the conversion of internal energy to kinetic and potential energy are possible, and the conclusion is inevitable that material bodies or systems possess an internal energy which in some way is connected with the measure of hotness or coldness of the body or system.

The existence of the internal energy associated with the material of a body can be deduced by experimentation in which the kinetic energy of the body or system is increased, and it is useful at this point to consider whether other forms of energy might be recognised by corresponding experiments in which the end result is an increase in kinetic energy. Figure 1.5.1 illustrates a number of experiments, including that already discussed, which can be used to demonstrate the existence of various energy forms.

It is now becoming clear that by deriving suitable experiments we can arrange for processes to take place which convert energy in its various forms into kinetic energy, which is easily observable. However, it is possible to arrange for conversion processes to take place between any energy forms, providing any necessary experimental complications are acceptable. In many processes which involve an energy transfer or conversion process, interchange takes place simultaneously between more than two energy forms, and it may not be possible to convert energy of one form directly into another without a third undesirable conversion taking place. It would be desirable for a nuclear power station to convert nuclear energy into electrical energy without increasing the internal energy of the environment, but such an outcome is not possible as will be shown later.

Any process involving energy conversion takes place subject to the law of conservation of energy which states that, for an isolated

Experimental system	Process and result	Energy form
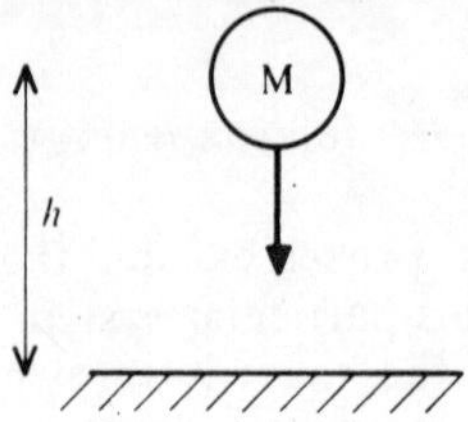	(k.e.) increases as h reduces	Gravitational potential energy
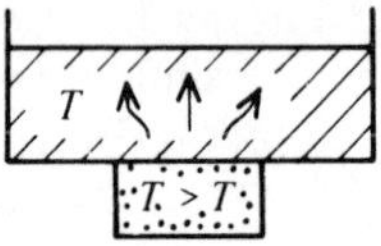	Convective motion of water produced by hot body but not by cold body	Internal energy
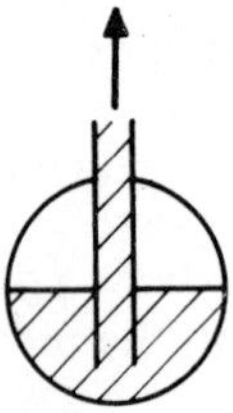	Chemical reaction produces increase of (k.e.) and potential energy	Chemical energy
	Two bodies with positive electrical charge repel and cause increase in (k.e.)	Electrical potential energy
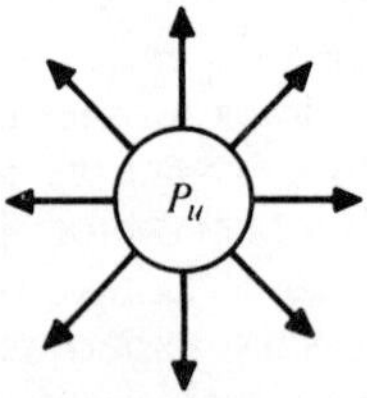	Atomic reaction results in explosive increase in (k.e.)	Nuclear energy

Fig. 1.5.1. Experiments in energy conversion.

system having no energy interaction with its environment, the total energy within the system remains constant. Energy interactions and conversions within the system are not prohibited, but they result in no change in the total energy contained within the system boundaries. Thus, if the total energy of the system at time t is $\mathcal{E}_t$ then for the isolated system the principle may be written

$$\mathcal{E}_{t_1} = \mathcal{E}_{t_2} = \mathcal{E}_{t_3} = \ldots \qquad [1.5.1]$$

As in the previous cases of conservation principles this is of little use unless it is known precisely what is meant by an isolated system and what happens if it is not. In the case of momentum transfer it was possible to find two identifiable methods of transfer which resulted in a momentum change and the case of energy transfer is very similar.

We take first the case of a mass M on the surface of the earth which is then raised slowly to a height h by applying a vertical force F. The force required to do this is given by $F = Mg$, where g is the gravitational acceleration. The potential energy it has acquired in this process is Mgh the product of the force Mg and the distance moved h. Thus if ϕ_1 is the initial potential energy and ϕ_2 the potential energy at height h then

$$\phi_2 - \phi_1 = Mgh \qquad [1.5.2]$$

The mass has now an increased potential energy as a result of a force doing work on it and the increase in potential energy *is* equal to the work done. Therefore the energy of a system can be increased by at least one identifiable process, that of *mechanical work* through the action of a force. The symbol generally used for energy transferred by a work process is W and it is taken to be positive for energy transfers *out* of the system.

Consider now a quite different situation where a mass at low temperature is brought into contact with a second mass at higher temperature. Common experience suggests that the result would be that the first mass would become hotter, in other words its internal energy would increase. No forces are involved in this interaction which cannot therefore be described as a work interaction. However, an energy transfer has taken place, and detailed experiment would confirm that its characteristic feature was that the presence of the temperature difference was a necessary requirement for the transfer to occur. This is called a *heat* interaction and the process taking place is conventionally called *heating* or *cooling*, depending on whether the internal energy is increasing or decreasing. Here, Q is the conventional symbol for the energy transfer resulting from a *heating* process. A positive value indicates an energy transfer *into* the system. Note that this is exactly the reverse of the convention for a work transfer.

All energy transfers across the boundary of a system are normally classed as heat or work, but to some extent this is confusing. A case in point is an electric current flow into a system. If the system is a simple resistor the result of energy transfer by the current is identical to that of heating. If the system is an electric motor the transfer of energy may more aptly be classed as a work effect, with perhaps a measure of heating as well. Similar problems arise when electromagnetic radiation to the system is considered. Radiation in the visible spectrum can produce motion of the vanes of a radiometer – a work effect – but if the vanes are held stationary the effect is that of heating. Photosynthesis in plants is produced by a radiative energy transfer which cannot be classed as either heat or work. Some caution is therefore necessary in special cases, but fortunately for our purposes here it will prove to be sufficient to distinguish only heat and work among energy-transfer processes and disregard other possibilities.

For our purposes, then, an isolated system may be defined as one having no mass transfer across the boundary and subjected to no energy transfer by work or heat. Under these circumstances the energy of the system will remain constant. It remains now to examine what happens when energy transfers to the system take place, and in particular the transfers that occur when a mass transfer takes place. We begin by considering the work which must be done when a mass transfer is made across the boundary of a system.

Entrance and exit work

We consider the system shown in Fig. 1.5.2 where a small plug of mass having cross-sectional area A, length δl and density ρ is about to be moved across the boundary and into the system. If the system is at pressure p then a force $F = pA$ must be exerted, otherwise the

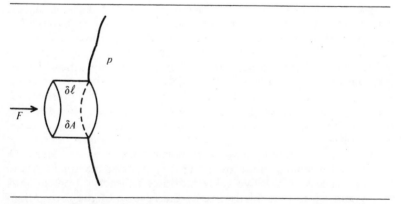

Fig. 1.5.2. Mass transfer into a system.

mass cannot possibly cross the boundary and enter the system. As the mass enters the system the work done by the force F in the process is the product $F\,\delta l$, and thus the process of mass transfer has required a work effect on the system, which now includes the small mass of magnitude $F\,\delta l$. Mass transfer across the boundary cannot take place without there being this corresponding energy transfer to the system during the process. If δV is the volume of the small element of mass then the element of work δW is given by

$$\delta W = F\,\delta l = pA\,\delta l = p\,\delta V \qquad [1.5.3]$$

or if W is the work done per unit mass then we have

$$W = p\left(\frac{\delta V}{\delta M}\right) = \frac{p}{\rho} = pv \qquad [1.5.4]$$

where v is the *specific volume* or inverse of the density. The product pv will be called the *entrance or exit work* per unit mass. Note it is a requirement for mass transfer to take place that the entrance work is done on the system, but in no sense is the product pv an energy property of the material itself.

Suppose now that the small mass δM has associated with it energy ε per unit mass. When the mass transfer across the boundary takes place the energy transfer to the system is the sum of the energy content of the mass itself and the entrance work and thus

$$\delta \mathcal{E} = \delta M(\varepsilon + pv) \qquad [1.5.5]$$

or

$$\delta \mathcal{E} = \delta M(u + pv + (\text{k.e.}) + \phi + \ldots) \qquad [1.5.6]$$

where u is the internal energy, (k.e.) the kinetic energy, and ϕ the potential energy, all taken on a per unit mass basis. The relationship of equation [1.5.6] completely specifies the energy transferred to a system by a mass flow across the boundary, and can now be used to form an equation expressing the principle of conservation of energy for a system which does have energy interaction with its environment.

In Fig. 1.5.3 the system of energy $\mathcal{E}$ has a work and heat interaction of ΔW and ΔQ respectively, and the elemental mass transfer is δM. Using the principle of conservation of energy, the energy balance for the system can be written, using equation [1.5.6],

$$\delta \mathcal{E} = \Delta Q - \Delta W + \delta M(u + pv + (\text{k.e.}) + \phi + \ldots) \qquad [1.5.7]$$

where $\delta \mathcal{E}$ is the change in total energy of the system contained within the boundary surface.

In processes where flow is taking place it is more convenient to use, instead of equation [1.5.7], an equation expressed in terms of energy-transfer rates and mass flow rates. The derivation of this

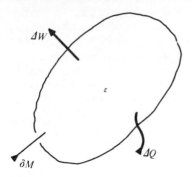

Fig. 1.5.3. Energy-transfer system.

from equation [1.5.7] is quite straightforward and yields

$$\frac{d\mathcal{E}}{dt} = \dot{Q} - \dot{W} + \dot{M}(u + pv + \text{(k.e.)} + \phi + \ldots) \qquad [1.5.8]$$

where $\dot{W}$, $\dot{Q}$ and $\dot{M}$ are respectively the rates of work, heat and mass transfers or fluxes across the boundary surface. For a system with a number of inlets and outlets the energy transfers resulting from them must be summed to give the overall energy balance. Using suffixes to indicate inlet and outlet parameters as before, we obtain for such a system

$$\frac{d\mathcal{E}}{dt} = \dot{Q} - \dot{W} + \sum_{\text{in}} \dot{M}(u + pv + \text{(k.e.)} + \phi + \ldots) -$$

$$- \sum_{\text{out}} \dot{M}(u + pv + \text{(k.e.)} + \phi + \ldots) \qquad [1.5.9]$$

If the system being considered does not have distinct inlets and outlets or conditions are not uniform across the cross-section, then, as in previous sections, an integration procedure must be adopted. The procedure, as before, is to take a small elemental surface δA having a mass flux $\delta \dot{M}$ across it and to write $\delta \dot{M} = \rho(\mathbf{v} . \delta \mathbf{A})$. The elemental mass flow $\delta \dot{M}$ is thus positive for flow into the system and negative for outward flow, consequently the distinction in equation [1.5.9] between inlet and outlet sections may now be dropped. The integral form of equation [1.5.9] thus becomes

$$\frac{d\mathcal{E}}{dt} = \dot{Q} - \dot{W} + \int_A \rho(\mathbf{v} . d\mathbf{A})(u + pv + \text{(k.e.)} + \phi + \ldots) \qquad [1.5.10]$$

For systems having variable energy and velocity parameters at the boundary surface it is essential that the correct integration procedure is adopted, i.e. that shown in equation [1.5.10], rather than some *ad hoc* averaging procedure used with either equations [1.5.8] or [1.5.9].

In many practical systems involving flow the energy of the system remains constant, i.e. $d\&/dt = 0$ and for these systems equations [1.5.9] and [1.5.10] simplify and may be written

$$\dot{W} - \dot{Q} = \sum_{in} \dot{M}(u + pv + (\text{k.e.}) + \phi + \dots) -$$
$$- \sum_{out} \dot{M}(u + pv + (\text{k.e.}) + \phi + \dots) \qquad [1.5.11]$$

$$\dot{W} - \dot{Q} = \int_{A} \rho(\mathbf{v}.\,d\mathbf{A})(u + pv + (\text{k.e.}) + \phi + \dots) \qquad [1.5.12]$$

In this form the equation in either presentation is known as the *steady-flow energy equation* and is found to be the basis of solution of a large number of practical problems in which the required conditions for validity are met.

An interesting development can now be made by applying equation [1.5.11] to the elemental stream tube for which a momentum balance was made in Section 1.4. The stream tube is shown again in Fig. 1.5.4.

If we make the assumption that the flow is steady and that the energy content of the element does not change with time, then the energy balance for the element is, from equation [1.5.11],

$$\dot{W} - \dot{Q} = \dot{M}(u + pv + (\text{k.e.}) + \phi + \dots) - \dot{M}[(u + \delta u) + (p + \delta p)(v + \delta v) +$$
$$+ ((\text{k.e.}) + \delta(\text{k.e.})) + (\phi + \delta\phi) + \dots] = 0$$

but $\dot{M}$ is the same at the inlet and outlet cross-sections and hence to the first order in small quantities for changes in internal, kinetic and potential energies only.

$$\dot{W} - \dot{Q} = -\dot{M}(\delta u + p\,\delta v + v\,\delta p + \delta(\text{k.e.}) + \delta\phi)$$

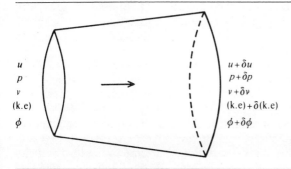

u	$u + \delta u$
p	$p + \delta p$
v	$v + \delta v$
(k.e)	(k.e.) + δ(k.e.)
ϕ	$\phi + \delta\phi$

Fig. 1.5.4. The elemental stream tube.

Dividing both sides of this equation by $\dot{M}$ and writing $\Delta Q = (\dot{Q}/\dot{M})$ and $\Delta W = (\dot{W}/\dot{M})$ we obtain

$$\Delta Q - \Delta W = \delta u + p\,\delta v + v\,\delta p + \delta(\text{k.e.}) + \delta\phi$$

$$= \delta u + p\,\delta v + v\,\delta p + \delta\left(\frac{v^2}{2}\right) + \delta\phi \qquad [1.5.13]$$

The corresponding equation for momentum balance was derived in section 1.4 and is equation [1.4.15], i.e.

$$\delta\left(\frac{v^2}{2}\right) + \frac{\delta p}{\rho} + \Delta\Phi + \Delta W + \delta\phi = 0$$

Replacing the inverse density by specific volume and rearranging, gives

$$v\,\delta p + \delta\left(\frac{v^2}{2}\right) + \delta\phi + \Delta W + \Delta\Phi = 0 \qquad [1.5.14]$$

which can now be combined with the energy equation [1.5.13] to give

$$\Delta Q = \delta u + p\,\delta v - \Delta\Phi \qquad [1.5.15]$$

This equation, a result of combining the equations representing both momentum and energy-conservation equations is known as the *equation of thermal interaction*.

The particular derivation given here shows that it is valid for steady-state flow along a stream tube where δu and δv are increments in internal energy and specific volume in the direction of flow, and where ΔQ, $\Delta\Phi$ are respectively the heating and frictional dissipation per unit mass associated with the element. Sometimes it is advantageous to look at the changes that occur within a particular element of fluid as it passes along a streamline, rather than to concentrate on changes between two cross-sections in a flow. The outcome of the two points of view are obviously related, and after a little consideration it is not difficult to see that equation [1.5.15] also represents the changes in internal energy, etc. that occur within a fluid element as it passes along the stream tube and equates these to the heating and dissipation within that particular element of fluid. Clearly u and v *do* vary with time for a particular element of fluid, and thus if equation [1.5.15] is applied to changes in an element of fluid as it proceeds through the flow, then neither the velocity nor any other parameter is required to be constant in time. However, it must be noted that if equation [1.5.15] is applied to changes between two cross-sections in a stream tube then, for it to be valid, it is necessary that there is no variation with time in the flow velocity and the total energy of the element of fluid between the two cross-sections.

1.6 Problems

1 In a hydraulic braking system the pistons in the operating cylinder and in the eight repeater cylinders have the same area of cross-section. The maximum travel of the operating piston is 28 mm. Find the corresponding distance travelled by the repeater pistons.

2 In a food-processing factory it is found that a vegetable soup tends to acquire an unpalatable appearance if passed through pipes at a velocity exceeding 1·5 m/sec. However, if the velocity is less than 1 m/sec the pipes become blocked with the solid material in the produce. The soup-canning machines can fill 2800 cans of soup each hour, each can containing 0·2 litres. Find a suitable diameter for the supply line to each machine.

3 In a typical oscilloscope tube the electron beam is 0·5 mm in diameter and the beam is directed to a particular spot on the screen by electrostatic deflector plates which are 15 mm long in the direction of the beam. If the electron velocity is 12×10^6 m/sec find the beam current and an upper limit to the frequency response of the oscilloscope. Take $\rho_c = 0·01$ C/m^3.

4 A rocky ledge 80 m below the crest of a waterfall is hit directly by the cascade of water. Find the pressure acting on the surface of the ledge due to the water. What is the velocity of the water just before it comes into the neighbourhood of the ledge?

5 A hydraulic crane is required to lift a load of 700 kg to a height of 8 m in 15 sec. Find the minimum power of hydraulic pump which will achieve this and, if the pressure rise of the pump is 10 000 kN/m^2, find the mass flow rate of fluid through the pump. Take $\rho = 850$ kg/m^3.

6 A domestic water tap is supplied from an open storage tank in the roof space and in the tank the water level is 6 m above the tap outlet. When the tap is fully open the rate of flow of water is 2 m^3/hr. Find the rate of frictional dissipation of energy when the tap is open and explain where the dissipation is taking place.

7 In a chemical process plant *n*-butyl alcohol, $\rho = 800$ kg/m^3, is pumped at the rate of 25 kg/sec from a storage tank where the pressure is 300 kN/m^2, to a reaction vessel at the same level where the pressure is 30 000 kN/m^2. Find the minimum pump power required to do this and give reasons why this is a minimum. Neglect changes in internal energy.

8 Water flows at a rate of 4 m^3/sec into a turbine through a supply pipe of diameter 0·4 m. The discharge pipe has a diameter of 1·2 m and the pressure drop across the turbine is 200 kN/m^2. Find the water velocity in the supply and discharge pipes and, assuming negligible change in internal and potential energy, determine the power output of the turbine.

Chapter 2

Material properties and constitutive equations

2.1 Introduction

In many practical problems it is necessary, when using the conservation laws, to have further understanding of the physical properties of substances and in particular of the factors which govern the transport of energy, momentum, charge and mass. It is common experience that a copper wire will readily allow an electric current flow while a nylon thread will not. In technical terms the nylon material is said to have a higher specific resistance or resistivity than copper. Resistivity thus enables comparisons between different materials to be made for the process of charge transport or electric current flow and is thus a material property. Other material properties include thermal conductivity, elasticity and viscosity – all having the feature that they allow comparisons to be made between materials for a particular type of transport process. There are many other classes of material property, but in this chapter it is the properties associated with transport processes that are the main concern.

There are many ways in which a material property for a particular process could be uniquely defined and then determined by simple experiment. In the case of resistivity this could be defined, for example, as the electric potential which needs to be applied across the ends of a filament of the material 1 m in length and 1 mm in diameter in order to produce a current of 1 mA. This is not the accepted definition, but nevertheless it is quite unambiguous and its value is readily measured for any material. Suppose it is required to select a wire diameter which will pass a current of 3 A when the potential difference over a 10 m length is 5 V. Does knowledge of the appropriate resistivity enable this to be done without any additional information? A little thought shows that since the diameter and length are different from those in the original experiment, the value is useless by itself. Apparently it is necessary to do further experiments to cover every conceivable diameter and length. The solution to this absurd situation is to use Ohm's law and, as will be shown later, by doing just that, one experiment on each material is quite sufficient. The object of this discussion is to show that material properties may prove of little value in practice unless they can be associated and used with a simple physical law, as for example Ohm's law in this particular case. Generally, material properties for transport processes have evolved from the laws governing that process in particular materials, and the material properties appear in the equations setting out the law in mathematical form. Thus, the property resistivity emerges as a physical constant in Ohm's law. Equations that relate material properties to the way a transport process proceeds in differing materials are known as *constitutive* equations. These equations and the material properties associated with them are the subject of this second chapter.

2.2 Thermal processes

The equations expressing the principle of conservation of energy developed in Chapter 1 are of little use unless suitable relationships are available which enable the internal energy of the material within a system to be found in terms of such measurable parameters as, for example, temperature and pressure. The objective in this section is to develop a number of thermodynamic relations that will enable not only values to be assigned to the internal energy of materials within systems but also to discuss a number of useful relationships between material properties including internal energy. Using these it will then prove possible to investigate a number of interesting processes, including that of phase change between solid, liquid and gas.

Internal energy

Consider first the process shown in Fig. 2.2.1 in which a substance contained in a closed, thermally insulated box is heated by means of an electrical heater. We assume that the box is sufficiently rigid so that its volume remains constant, and note that in this example the electric current is providing a heating effect rather than a work effect. We take ΔQ to be the heat input per unit mass of substance over some time period for which δu and δT are the corresponding

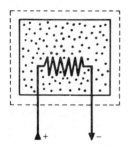

Fig. 2.2.1. Constant-volume heating process.

increases in internal energy and temperature. From equation [1.5.15] we have

$$\Delta Q = \delta u + p\,\delta v - \Delta\Phi$$
$$= \delta u \qquad\qquad [2.2.1]$$

since $\delta v = 0$ and there need be no motion to produce any surface frictional effects. The specific heat of substance at constant volume c_v is by definition given by

$$c_v = \left(\frac{\Delta Q}{\delta T}\right)_{\text{constant volume}}$$

and hence

$$\Delta Q = c_v\,\delta T \qquad\qquad [2.2.2]$$

Combining equations [2.2.1] and [2.2.2] gives finally

$$\delta u = c_v\,\delta T \qquad\qquad [2.2.3]$$

and upon integration we find for a finite process between temperatures T_1 and T_2 the corresponding change in internal energy from u_1 to u_2 is

$$u_2 - u_1 = \int_{T_1}^{T_2} c_v\,\mathrm{d}T \qquad\qquad [2.2.4]$$

We have thus determined an expression for internal energy change in terms of temperature, and the specific heat at constant volume which can readily be determined by experiment and exists for most materials in tabulated form. Absolute values of internal energy are rarely required, and as a result the datum level at which the internal energy is taken to be zero is not of great practical interest.

Enthalpy

Of Greek derivation, the name *enthalpy* is given to the sum of the internal energy and the product pv. Thus the enthalpy h is by definition given by

$$h = (u + pv) \qquad\qquad [2.2.5]$$

Interest in this quantity stems from the energy equations [1.5.8], [1.5.9] or [1.5.10] where the terms $(u + pv)$ are in each case present on the right-hand side. A further example will assist in understanding why this is a useful quantity to use, rather than the separate terms themselves.

We consider in this case the heating process illustrated in Fig. 2.2.2. Here again an electrical heater is used to heat a substance in

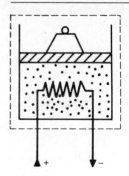

Fig. 2.2.2. Constant-pressure heating process.

an insulated box, but now the top of the box is an insulated weighted piston which allows the substance to expand freely at a constant pressure. When the heater is switched on, the process is a heating at *constant pressure* rather than constant volume as in the previous example. We now apply equation [1.5.15] to this process, assuming that the process is slow so that frictional forces are negligible. Thus

$$\Delta Q = \delta u + p\,\delta v - \Delta\Phi$$
$$= \delta u + p\,\delta v$$
$$= \delta u + p\,\delta v + v\,\delta p$$

since $\delta p = 0$, or

$$\Delta Q = \delta u + \delta(pv) = \delta h \qquad [2.2.6]$$

However, for a constant-pressure heating process we know that

$$\Delta Q = c_p\,\delta T \qquad [2.2.7]$$

and hence from equations [2.2.6] and [2.2.7] we finally obtain

$$\delta h = c_p\,\delta T \qquad [2.2.8]$$

which upon integration between temperatures T_1 and T_2 gives

$$h_2 - h_1 = \int_{T_1}^{T_2} c_p\,dT \qquad [2.2.9]$$

This equation is most significant, for not only does it show that enthalpy changes can be expressed as a function of specific heat and temperature change, as in the case of internal energy u, but we can go further. We note that

$$\delta h = \delta u + \delta(pv)$$

or, using equations [2.2.8] and [2.2.3],

$$c_p \, \delta T = c_v \, \delta T + \delta(pv)$$

i.e.

$$(c_p - c_v)\delta T = \delta(pv) \tag{2.2.10}$$

This latter equation is a relationship between δT and $\delta(pv)$ which appears to be true for any chosen substance since nowhere in the analysis was any particular material selected. It suggests quite strongly that for a selected substance there exists some relationship between the pressure, specific volume and temperature at which that substance can exist. In other words for any substance p, v and T are not independent properties but, given any two, the value of the third is predetermined by the nature of the material itself. Thus

$$p = p(v \; T), \qquad v = v(T, p), \qquad T = T(p, v) \tag{2.2.11}$$

the actual functional relationships depending on the material itself. Furthermore, we have $\delta u = c_v \, \delta T$ and $\delta h = c_p \, \delta T$, and consequently we find equation [2.2.11] suggests that

$$u = u(p, v) \quad \text{or} \quad u = u(v, T) \quad \text{or} \quad u = u(T, p) \tag{2.2.12}$$

and

$$h = h(p, v) \quad \text{or} \quad h = h(v, T) \quad \text{or} \quad h = h(T, p) \tag{2.2.13}$$

The parameters p, v, T, u and h are examples of what are called *thermodynamic properties*, and in fact the notions suggested by equations [2.2.11], [2.2.12] and [2.2.13] are found to be correct, namely, that for a given substance only two properties are completely independent and any other is a functional relationship of the two. Only in special cases can the interrelationships be found theoretically. This law of nature is generally called the *two-property rule*.

The perfect gas

A perfect gas by definition is one for which the pressure, temperature and specific volume satisfy the following relationship:

$$pv = RT \tag{2.2.14}$$

where R is the *gas constant* for the gas. The law expressed in equation [2.2.14] is known as the *perfect gas law* and is found to be accurate for most gases of low molecular weight, providing the temperature and pressure are such that the gas is not close to conditions where it may be liquefied. The perfect gas law is an extremely simple example of the two-property rule. In the form in which it is written in equation [2.2.14] the value of the gas constant

R depends on the gas involved, but it turns out that the value of R is solely dependent on the molecular weight of the gas and on no other parameter. The appropriate equation for R is found to be

$$R = \frac{R_0}{m} \qquad [2.2.15]$$

where R_0 is the universal gas constant having a value of 8·3143 kJ/mol K and m is the molecular weight in kilograms. Thus, for oxygen having a molecular weight of 32 the gas constant is

$$R = \frac{8 \cdot 3143}{32} = 0 \cdot 259\,82\,\text{kJ/kg K}.$$

For a perfect gas there is an interesting and often very useful relationship between the specific heats c_v, c_p and the gas constant R. The key to this is equation [2.2.10], i.e.

$$(c_p - c_v)\,\delta T = \delta(pv)$$

but from the perfect gas law

$$\delta(pv) = R\,\delta T$$

and thus for a perfect gas

$$(c_p - c_v)\,\delta T = R\,\delta T$$

or

$$(c_p - c_v) = R = \frac{R_0}{m} \qquad [2.2.16]$$

Thus the gas constant is equal to the difference between the specific heats at constant pressure and constant volume. For a perfect gas, knowledge of the molecular weight is sufficient to determine both the gas constant R and the difference in specific heats $(c_p - c_v)$.

It is appropriate, before leaving the study of a perfect gas for the moment, to consider a particular thermodynamic process which might take place in a perfect gas. The particular process is one in which the heat transfer $\Delta Q = 0$, i.e. *adiabatic* and also the frictional dissipation $\Delta \Phi = 0$. Such a process is ideal in the sense that for all practical processes $\Delta \Phi \neq 0$, but many processes approximate to this condition. With $\Delta Q = 0$ and $\Delta \Phi = 0$ equation [1.5.15] becomes

$$0 = \delta u + p\,\delta v$$
$$= c_v\,\delta T + p\,\delta v \qquad [2.2.17]$$

However, for a perfect gas

$$pv = RT$$

or

$$\frac{(p\,\delta v + v\,\delta p)}{R} = \delta T$$

Substituting in equation [2.2.17] for δT gives

$$\frac{c_v}{R}(p\,\delta v + v\,\delta p) + p\,\delta v = 0$$

Therefore

$$(c_v + R)p\,\delta v + c_v\,\delta p = 0$$

but $(c_v + R) = c_p$ for a perfect gas and hence

$$\left(\frac{c_p}{c_v}\right)p\,\delta v + v\,\delta p = 0$$

and thus p and v satisfy the following relationship:

$$pv^{(c_p/c_v)} = \text{constant}$$

The ratio (c_p/c_v) is conventionally given the symbol γ, and thus for an adiabatic frictionless process in a perfect gas the pressure and specific volume are related by the equation

$$pv^\gamma = \text{constant} \qquad [2.2.18]$$

where $\gamma = (c_p/c_v)$.

It is now straightforward to use the perfect gas law together with equation [2.2.18] to develop corresponding relationships between p, T and v, T. These are

$$pT^{-\gamma/(\gamma-1)} = \text{constant} \qquad [2.2.19]$$

$$Tv^{\gamma-1} = \text{constant} \qquad [2.2.20]$$

Entropy

The thermodynamic properties p, v, T, u and h so far introduced by no means represent a complete set, and further properties, obeying the two-property rule, can readily be developed. Perhaps the most useful of these is the property entropy usually denoted by the symbol s and defined by the following equation

$$T\,\delta s = \delta u + p\,\delta v \qquad [2.2.21]$$

This equation defines *changes* in entropy rather than absolute values, but for our present purposes it will not be necessary to go further and discuss the selection of datum values. Recalling equation [1.5.15] we have

$$\Delta Q = \delta u + p\,\delta v - \Delta \Phi$$

and using equation [2.2.21] we obtain

$$T\,\delta s = \Delta Q + \Delta \Phi \qquad [2.2.22]$$

It can be seen now that the rather special process where $\Delta Q = 0$ and $\Delta \Phi = 0$ has the significant feature that $\delta s = 0$, and it is therefore a process without change in entropy. A process which occurs without change in entropy is formally called *isentropic* and, as equation [2.2.22] shows, an example of an isentropic process is one which is both adiabatic and frictionless.

For flow processes without heating or cooling, i.e. $\Delta Q = 0$, equation [2.2.22] reduces to the simpler form

$$T\,\delta s = \Delta \Phi \quad \text{or} \quad \delta s = \frac{\Delta \Phi}{T}$$

and from this it can be seen that *changes* in entropy are in this case a measure of the frictional dissipation that is occuring in the flow. We shall be returning to this point later in detail, but even now it is useful to note this possibility.

Phase change

The thermodynamic properties of substances are closely related to the degree of packing of the molecules, and for substances which can exist in different phases – i.e. as a solid, liquid or gas – significant changes take place in the property values during the process of phase change. Figure 2.2.3 illustrates what happens typically if a solid substance is heated at constant pressure until it first changes phase, or melts into a liquid and then, after further heating, enters the gaseous phase. On this diagram the horizontal coordinate is enthalpy and the vertical, temperature. The important point to note

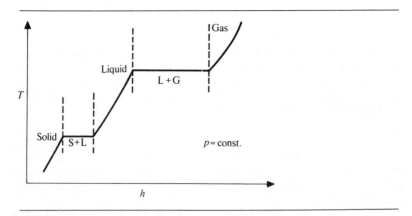

Fig. 2.2.3. The phase-change process.

is that during each of the phase-change processes where a mixture of two phases exist together the temperature remains constant. The total change in enthalpy during either of these processes of phase change is equal to the appropriate latent heat, for we have from equation [1.5.15] with $\Delta\Phi = 0$

$$\Delta Q = \delta u + p\,\delta v$$

but p = constant, therefore

$$\Delta Q = \delta u + \delta(pv)$$
$$= \delta h$$

and hence over the phase-change process the difference in enthalpy is equal to the energy supplied by heating, in other words the latent heat.

There are many ways to illustrate – in diagram form – changes in thermodynamic properties during phase change, but one that will be found most useful later is by plotting temperature and entropy as shown in Fig. 2.2.4 for the liquid-to-vapour process.

In Fig. 2.2.4 a set of curves is drawn, each representing the relationship between T and s for a constant-pressure change of phase. The *saturation curve* shown on the diagram is drawn through the points at the beginning and end of each phase-change process at the various pressures. The maximum value of T on the curve is called the *critical point* and the corresponding temperature and pressure are known as the *critical temperature and pressure* respectively. At pressures above the critical value the substance fails to be adequately described as either liquid or gas and does not change phase in the usual sense.

It is often necessary to determine values of u, h and s when the substance has values of p and T which lie within the saturation curve of Fig. 2.2.4. In order to do this we require an additional

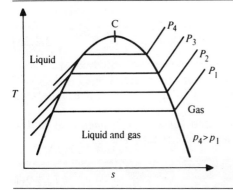

Fig. 2.2.4. The temperature/entropy or Ts diagram.

parameter as well as p or T to determine property values. The usual parameter for this is known as the *dryness fraction* and is given the symbol x. The dryness fraction is defined as being the proportion of substance by mass which exists in vapour or gaseous form. Thus, if we take a mass M of the substance and of this M_g is the mass of the gaseous portion of the mixture then $x = M_g/M$ and clearly $(1-x)$ is the corresponding proportion of liquid in the mixture. It is now straightforward to determine the corresponding values for the properties v, u, h and s, providing appropriate values are known for the properties when $x = 0$ and $x = 1$, i.e. the values on the saturation curve. It is usual to denote these particular values by a suffix notation, thus

u_l = internal energy when $x = 0$, or when all the substance is just liquid.
u_g = internal energy when $x = 1$, or when all the substance is just gaseous.

An obvious notation is used for the difference $(u_g - u_l)$ which is written u_{lg}. The total internal energy of a mixture is the sum of the energies of the two constituents and in terms of the internal energy per unit mass we have for the mixture

$$Mu = M_l u_l + M_g u_g$$

Therefore

$$u = \left(\frac{M_l}{M}\right)u + \left(\frac{M_g}{M}\right)u_g$$

i.e.

$$u = (1-x)u_l + xu_g \qquad [2.2.23]$$

or

$$u = u_l + xu_{lg} \qquad [2.2.24]$$

or

$$u = u_g - (1-x)u_{lg} \qquad [2.2.25]$$

Common practice is to use whichever of equations [2.2.23]–[2.2.25] proves to be easier and more accurate to use, and this to some extent depends on the particular way in which tabulated values of u_l, u_g and u_{lg} are presented.

Corresponding relationships hold for the properties v, h and s. Written in the form of equation [2.2.23] they are

$$v = (1-x)v_l + xv_g$$
$$h = (1-x)h_l + xh_g \qquad [2.2.26]$$
$$s = (1-x)s_l + xs_g$$

p	T	u_1	u_g	h_1	h_{1g}	h_g	s_1	s_{1g}	s_g
kN/m² °C		kJ/kg		kJ/kg			kJ/kg K		
42	77·1	323	2478	323	2315	2638	1·040	6·612	7·652
44	78·2	327	2479	327	2313	2640	1·054	6·582	7·636
46	79·3	332	2481	332	2310	2642	1·067	6·554	7·621
48	80·3	336	2482	336	2308	2644	1·079	6·528	7·607
50	81·3	340	2483	340	2305	2645	1·091	6·502	7·593

Fig. 2.2.5. Saturated water and steam values.

It is left as an exercise for the reader to prove these relationships. In proving the second and third it may be found useful to invoke equation [1.5.15] and for the entropy also equation [2.2.21].

The manner in which the tabulated values are typically presented is shown if Fig. 2.2.5 and is fairly self-explanatory. The pressure and temperature on the left are the saturation values and may be presented with either p or T taking the integral values.

Tables also play an essential role in determining property values outside the saturation curve, particularly for gas close to the saturation curve where in most cases the perfect gas law is completely invalid and produces gross inaccuracy. The area outside and to the right of the saturation curve in Fig. 2.2.4 is known as the *superheat region*, and a gaseous substance whose pressure and temperature lie in this region is said to be *superheated*. Tabulated values in this region are typically presented as shown in Fig. 2.2.6.

In the extreme left column T_{sat} is the saturation temperature which would have prevailed had the gas been at saturation condition at the selected pressure p. The next column gives values for the saturation properties u, h and s. Columns to the right give corresponding values for various temperatures in the superheated region above the saturated value T_{sat}. Interpolation between values of properties in two columns is necessary for intermediate temperatures.

	T °C							
$p = 800$ kN/m²	200	250	300	350	400	450	500	550
T_{sat} 170·4 °C u_{sat} 2577	2631	2716	2798	2878	2960	3042	3126	3298	u kJ/kg
T_{sat} 170·4 °C h_{sat} 2769	2840	2951	3057	3162	3267	3373	3481	3699	h kJ/kg
T_{sat} 170·4 °C s_{sat} 6.663	6·817	7·040	7·233	7·409	7·571	7·723	7·866	8·132	s kJ/kg K

Fig. 2.2.6. Superheated steam values.

2.3 Thermal conduction

The process of thermal conduction, whereby a temperature difference produces an energy transfer from the hotter region to the cooler, is an everyday experience that needs no introduction. It is part of this experience that the rate at which this process takes place depends on the nature of the material itself. Hot coffee in a vacuum flask loses internal energy less rapidly than it would in a simple glass bottle. In cold weather, houses with thermal insulation require less fuel to retain suitable internal temperatures than those with none. The term 'thermal insulation' seems so similar to the term 'electrical insulation' that it is tempting to seek to explain matters by using an analogy with electrical phenomena. There it is known that the electric current is proportional to the potential gradient and another quantity, which is a property of the material. In thermal conduction this is, in fact, the way materials behave and the energy flux *is* found to be proportional to the product of a potential gradient, the temperature gradient, and a material property called the thermal conductivity. Therefore, if $\dot{q}$ is the rate of energy flow per unit area and dT/dx the temperature gradient normal to the area and in the same sense as $\dot{q}$, then

$$\dot{q} = -\kappa \frac{dT}{dx} \qquad\qquad [2.3.1]$$

where κ is called the thermal conductivity and depends on the nature of the material in which the temperature gradient exists. It follows from equation [2.3.1] that if $dT/dx = 0$ then $\dot{q} = 0$ and the energy flux is zero. If the temperature gradient is reversed in direction then so also is $\dot{q}$. It is most important to note that the direction of energy flow is in the direction of *falling* temperature. The law expressed in equation [2.3.1] is known as Fourier's law and is valid for most homogeneous and isotropic materials. It may prove to be seriously in error for non-isotropic materials such as resin-bonded glass-fibre materials. Values of the thermal conductivity for common materials can vary enormously – for copper $\kappa = 390$ W/m K and in contrast $\kappa = 0.17$ W/m K for asbestos. A table of values for common materials is given in Fig. 2.3.1.

Of particular interest in solid materials is the steady-state heat-conduction process in three dimensions. Such a process is, by definition, one in which the temperature at a point in the solid is a function of the coordinates of the point but not of time. The process can be analysed using equation [2.3.1] and considering its application to the elementary cube of material shown in Fig. 2.3.2. In this diagram the temperature and heating are given for the pair of faces having normals in the direction OX and for clarity the corresponding values are omitted from the remaining faces of the cube. Taking

T K	250	300	400	500	600	800
Aluminium	208	202	209	222	234	277
Copper	393	386	377	372	367	357
Steel	57	55	52	48	45	38
Glass		0·8–1·1				
Concrete		0·9–1·4				
Asbestos		0·163	0·194	0.206		

Fig. 2.3.1. Thermal conductivity values in W/m K.

this first pair as an example, the net flow of energy *into* the cube from these surfaces, say $\dot{\mathcal{E}}_x$, is given by

$$\dot{\mathcal{E}}_x = \dot{q}_x(\delta y\,\delta z) - (\dot{q}_x + \delta\dot{q}_x)(\delta y\,\delta z)$$

$$= \frac{\partial \dot{q}_x}{\partial x}(\delta x\,\delta y\,\delta z)$$

$$= \frac{\partial}{\partial x}\left(-\kappa\frac{\partial T}{\partial x}\right)(\delta x\,\delta y\,\delta z)$$

$$= -\kappa\frac{\partial^2 T}{\partial^2 x^2}(\delta x\,\delta y\,\delta z)$$

It follows that the corresponding contributions from the heat fluxes in the OY and OZ directions, respectively, are

$$\dot{\mathcal{E}}_y = -\kappa\frac{\partial^2 T}{\partial y^2}(\delta x\,\delta y\,\delta z) \quad \text{and} \quad \dot{\mathcal{E}}_z = -\kappa\frac{\partial^2 T}{\partial z^2}(\delta x\,\delta y\,\delta z)$$

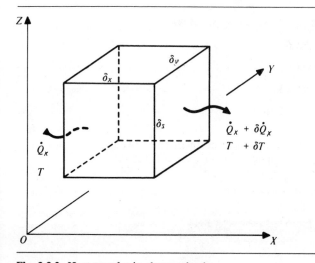

Fig. 2.3.2. Heat transfer in elemental cube.

However, the net energy flux into the cube must be zero if the system is in a steady state, and thus

$$\dot{\mathcal{E}}_x + \dot{\mathcal{E}}_y + \dot{\mathcal{E}}_z = 0$$

or

$$-\kappa\left(\frac{\partial^2 T}{\partial x^2} + \frac{\partial^2 T}{\partial y^2} + \frac{\partial^2 T}{\partial z^2}\right)(\delta x\, \delta y\, \delta z) = 0$$

The temperature T for steady-state heat conduction therefore satisfies the following differential equation

$$\frac{\partial^2 T}{\partial x^2} + \frac{\partial^2 T}{\partial y^2} + \frac{\partial^2 T}{\partial z^2} = 0 \qquad\qquad [2.3.2]$$

Example 2.3.1
A large area plate of material of thickness 10 mm has temperatures of 400 K and 300 K maintained at each of its surfaces. Find the rate of energy flow per square metre through the slab assuming (a) the material is steel, and (b) the material is asbestos. Find the energy flow and interface temperature if the two plates are placed together and the same temperatures are maintained on the outer surfaces.

Data
$$T_1 = 400 \text{ K}, \qquad T_2 = 300 \text{ K}$$
$$(x_2 - x_1) = 0{\cdot}01 \text{ m}$$

From Table 2.3.1 we have for steel,

$$\kappa_s = 53{\cdot}5 \text{ W/m K}$$

and for asbestos,

$$\kappa_a = 0{\cdot}178 \text{ W/m K}$$

Analysis
For steady-state heat conduction in the OX direction only, equation [2.3.2] gives with $dT/dy = dT/dz = 0$, that

$$\frac{d^2 T}{dx^2} = 0 \quad \text{or} \quad \frac{dT}{dx} = \text{constant}$$

Therefore, from equation [2.3.1]

$$\dot{q} = -\kappa \frac{dT}{dx} = -\kappa \frac{(T_2 - T_1)}{(x_2 - x_1)}$$

Calculation

For (a):

$$\dot{q}_s = -\frac{53 \cdot 5 \times (400 - 300)}{0 \cdot 01}$$

$$= -535 \times 10^3 \text{ W/m}^2$$

$$= \mathbf{-535 \text{ kW/m}^2}$$

For (b):

$$\dot{q}_a = -\frac{0 \cdot 178 \times (400 - 300)}{0 \cdot 01}$$

$$= -1 \cdot 780 \times 10^3 \text{ W/m}^2$$

$$= \mathbf{-1 \cdot 78 \text{ kW/m}^2}$$

Analysis

Let T_i be temperature at interface between steel and asbestos plates. Then for steel plate we have

$$\dot{q}_s = -\kappa_s \frac{(T_1 - T_i)}{(x_2 - x_1)}$$

and for asbestos plate

$$\dot{q}_a = -\kappa_a \frac{(T_i - T_2)}{(x_2 - x_1)}$$

However, the principle of conservation gives $\dot{q}_s = \dot{q}_a$, and hence

$$\kappa_s(T_1 - T_i) = \kappa_a(T_i - T_2)$$

or

$$T_i(\kappa_s + \kappa_a) = (\kappa_s T_1 + \kappa_a T_2)$$

Therefore

$$T_i = \frac{(\kappa_s T_1 + \kappa_a T_2)}{(\kappa_s + \kappa_a)}$$

Calculation

$$T_i = \frac{(53 \cdot 5 \times 400 + 0 \cdot 178 \times 300)}{(53 \cdot 5 + 0 \cdot 178)}$$

$$= \mathbf{399 \cdot 7 \text{ K}}$$

$$\dot{q} = -\frac{0 \cdot 178(399 \cdot 7 - 300)}{0 \cdot 01}$$

$$= 1 \cdot 77 \times 10^3 \text{ W/m}^2$$

$$= \mathbf{1 \cdot 77 \text{ kW/m}^2}$$

2.4 Viscosity and frictional dissipation

In section 1.4 we introduced the force f_s per unit mass which was taken to be the force resulting from the frictional interaction between elements of fluid. We now investigate the nature of this surface force and begin by considering the deformation experienced by a cubic element of fluid in the presence of a velocity gradient as shown in Fig. 2.4.1. In this diagram we consider particularly the face of area $(\delta x \, \delta y)$ and assume that the velocity gradient is du/dy.

The two parts of this diagram illustrate that if at time $t_1 = 0$ the face is rectangular the effect of the velocity gradient has been to cause a deformation in its shape to that of a parallelogram in time $(t_2 - t_1)$. It is this physical rate of deformation that results in the fluid exerting a surface force or shear stress on the surrounding elements in such a direction as to appear to resist the deformation. Much the same process occurs in elastic deformation of a cube of rubber, but a fluid is rather different in that it cannot return by itself to the original shape. In a fluid it is found that the important parameter is the *rate* of deformation, rather than the actual deformation as in an elastic solid.

In Fig. 2.4.1 the displacement in position of point 2 relative to point 1 is, over the time t_2, equal to $(\delta u)t_2$ and thus the rate of displacement of point 2 is just δu. However, the value of δu is not sufficient to define a rate of deformation since it depends, in fact is proportional to, the side δy of the selected element. This can be resolved completely if the rate of deformation is defined as $\delta u/\delta y$, or in the limit as the velocity gradient du/dy. All ambiguity is thus removed and the rate of deformation is completely independent of the size of the element being considered.

Experiments in many simple fluids produce an important relationship between the rate of deformation and the surface force tending to resist deformation. The relationship is known as *Newton's equation* and is, for the cube of Fig. 2.4.1,

$$\tau_x = \mu \frac{du}{dy} \qquad\qquad [2.4.1]$$

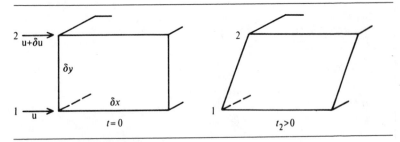

Fig. 2.4.1. Deformation in a fluid.

where τ_x is the force per unit area, or *shear stress* acting in the direction required to oppose the deformation. The coefficient μ is a property of the fluid and is called the *viscosity*. The simple expression of equation [2.4.1] is strictly valid only for flows in which changes in density can be ignored and, even then, only when the velocity gradient is solely in the one direction, in this case the OY direction. In the general case a more complicated expression must be used for evaluating shear stresses, but this need not concern us here.

It is interesting now to develop what is called an *equation of motion* for a small element of viscous fluid flowing under the action of a velocity gradient, du/dy. Consider the element of fluid shown in Fig. 2.4.2 for a steady fluid flow having all velocity components zero except for the X component, and du/dy the only non-zero velocity gradient. We ignore for the moment volume forces such as those due to gravitation or an electric potential. The net force acting in the OX direction on this element is

$$p(\delta y\, \delta z) - (p + \delta p)(\delta y\, \delta z) + (\tau + \delta\tau)(\delta z\, \delta x) - \tau(\delta z\, \delta x)$$

$$= -\delta p(\delta y\, \delta z) + \delta\tau(\delta z\, \delta x)$$

which to the first order in small quantities is

$$\left(-\frac{dp}{dx} + \frac{d\tau}{dy}\right)(\delta x\, \delta y\, \delta z)$$

If the flow is steady then the distance moved by the particle in time δt is u δt. Over this time period the particle velocity has changed by u δt(du/dx) and thus the acceleration of the particle in a steady flow is

$$\text{u } \delta t\frac{du}{dx} \cdot \frac{1}{\delta t} \quad \text{or} \quad \text{u } \frac{du}{dx}$$

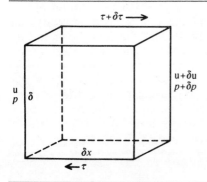

Fig. 2.4.2. An element of fluid.

The equation of motion can now be written down and is

$$\rho(\delta x\,\delta y\,\delta z)\,u\frac{du}{dx} = \left(-\frac{dp}{dx}+\frac{d\tau}{dy}\right)(\delta x\,\delta y\,\delta z)$$

or

$$u\frac{du}{dx} = -\frac{1}{\rho}\frac{dp}{dx}+\frac{1}{\rho}\frac{d\tau}{dy}$$

However, $\tau = \mu\,(du/dy)$ and we finally obtain

$$u\frac{du}{dx} = -\frac{1}{\rho}\frac{dp}{dx}+\frac{\mu}{\rho}\frac{d^2u}{dy^2} \qquad [2.4.2]$$

It is interesting to compare this result with equation [1.4.15] with the potential term taken to be zero. This equation may then be written

$$\delta\left(\frac{u^2}{2}\right)+\frac{\delta p}{\rho}+\Delta\Phi+\Delta W = 0$$

or writing in differential form and rearranging

$$u\frac{du}{dx} = -\frac{1}{\rho}\frac{dp}{dx}-\dot{\Phi}-\dot{W} \qquad [2.4.3]$$

where now $\dot{\Phi}$ and $\dot{W}$ are the rates of dissipation and working respectively.

Comparison of equations [2.4.2] and [2.4.3] shows that for this flow we have

$$\dot{\Phi}+\dot{W} = -\frac{\mu}{\rho}\frac{d^2u}{dy^2} \qquad [2.4.4]$$

and the form that the rate of dissipation and working terms may take is thus displayed clearly. However, equation [2.4.4] does *not* allow the separate evaluation of $\dot{\Phi}$ and $\dot{W}$. The evaluation of the rate of dissipation can, however, be made in terms of viscosity and the velocity gradient by considering the fluid element shown in Fig. 2.4.3. For this element we wish to evaluate the work done by the shear stresses in producing deformation of the element and to do this we take a set of axes OXYZ that are fixed in the base of the element as shown. The rate of working against the fluid to produce the deformation is thus the product of the force acting on the upper surface and the velocity of that surface. Thus, if $\dot{\Phi}$ is the rate of working per unit mass of fluid then

$$\dot{\Phi}\rho(\delta x\,\delta y\,\delta z) = \delta u\tau_x(\delta z\,\delta x)$$

$$= \frac{du}{dy}\,\tau_x(\delta x\,\delta y\,\delta z)$$

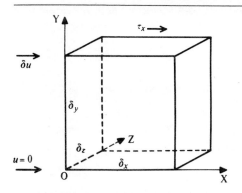

Fig. 2.4.3. Stress on a fluid element.

or in other words

$$\dot{\Phi} = \frac{1}{\rho}\,\tau_x\,\frac{du}{dy}$$

If the fluid is a *Newtonian fluid*, i.e. satisfies equation [2.4.1], then this may be written

$$\dot{\Phi} = \frac{\mu}{\rho}\left(\frac{du}{dy}\right)^2 \qquad\qquad [2.4.5]$$

and after using this result to eliminate $\dot{\Phi}$ from equation [2.4.4] we obtain the corresponding result for $\dot{W}$ which is

$$\dot{W} = -\frac{\mu}{\rho}\left\{\left(\frac{du}{dy}\right)^2 + \frac{d^2u}{dy^2}\right\} \qquad\qquad [2.4.6]$$

Some interesting observations arise from these two results. Firstly $\dot{\Phi}$ is necessarily positive or zero regardless of the flow details, and on the contrary $\dot{W}$ can take either positive or negative values depending on the relationship between $(du/dy)^2$ and d^2u/dy^2. The former observation can be taken a little further by considering the implication of this in relationship to the equation of thermal interaction, equation [1.5.15] and the definition of entropy equation [2.2.21]. We have from these

$$\Delta Q = \delta u + p\,\delta v - \Delta\Phi$$

and

$$T\,\delta s = \delta u + p\,\delta v$$

giving with $\Delta Q = 0$ for the process being considered

$$0 = T\,\delta s - \Delta\Phi$$

which may now be written as a rate equation for the fluid element in the form

$$T\frac{ds}{dt} = \dot{\Phi}$$
[2.4.7]

Using equation [2.4.5] to evaluate $\dot{\Phi}$ gives finally

$$\frac{ds}{dt} = \frac{\mu}{\rho T}\left(\frac{du}{dy}\right)^2 \geqslant 0$$
[2.4.8]

Thus the effect of fluid viscosity is that the entropy of a fluid element always *increases* under the action of a velocity gradient.

2.5 Molecular mixing processes

The phenomena we have been dealing with until now have all involved processes taking place in a single chemical substance, or in a uniform mixture of substances. However, in many of our everyday experiences we are involved in the mixing of different chemical species where the concentration of the species can vary from point to point. Smoke leaving a chimney and diffusing into the atmosphere is an example of a mixing process in which the concentration of the smoke, highest at the point of emission, varies with distance from the chimney. A small quantity of table salt placed in a beaker of water eventually dissolves and then diffuses through the water until all the water is effectively at the same degree of saltiness. This diffusion of the salt molecules among the water molecules is the result of random motion of the molecules and takes place without any visible sign of motion of the liquid. However, the process of mixing can be accelerated by agitation – as happens when sugar is stirred into coffee. The latter process is known as *convective mass transfer*, and the diffusion in a quiescent fluid, solely the consequence of molecular motion, is called *molecular mass transfer*. Convective mass transfer is of immense practical importance, but its study is outside the scope of the work here and we shall confine ourselves to a study of the molecular diffusion process.

We consider, then, a fluid having two identifiable components which we label A and B and in which the relative proportions of A and B vary from point to point. There are a number of ways in which we can define a parameter which identifies the relative properties of the two components. An obvious method is to define a *mass concentration*, ρ_A, as being the mass of component A present in unit volume of the mixture with a corresponding definition for ρ_B. The density of the mixture ρ is then given by

$$\rho = \rho_A + \rho_B$$
[2.5.1]

However, since we are dealing with a process which is the result of molecular motion and which depends in some way on the relative freedom of molecules to move randomly among each other, it turns out to be more useful generally to use a parameter known as the *molar concentration* c_A, defined as the number of moles of substance A present in unit volume of the mixture. A mole of any substance is that quantity that has a mass equal in numerical magnitude to its molecular weight and thus, for example, a mole of oxygen, O_2, is 32 kg of oxygen. The mass and molar concentrations are related to the molecular weight m by

$$\rho_A = c_A m_A, \qquad \rho_B = c_B m_B \qquad\qquad [2.5.2]$$

and thus from equation [2.5.1] we have

$$\rho = c_A m_A + c_B m_B \qquad\qquad [2.5.3]$$

Either of the parameters ρ_A and c_A can now be used to determine the relative proportions of A and B in any region of the fluid, and what is now required is corresponding parameters that show the amount of mass that is diffusing across any selected surface in the fluid. Our previous work in section 1.2 tends to suggest that a suitable parameter could be found by defining a *mass flux* j_A as being the rate at which mass of component A was being diffused or transported across unit area in the fluid. Similarly, a *molar flux* J_A can be defined as the number of *moles* of A that are diffused through the unit area in unit time. Again, these two alternative parameters are related by a simple equation, namely

$$j_A = J_A m_A, \qquad j_B = J_B m_B \qquad\qquad [2.5.4]$$

In section 2.3 we showed that the thermal flux was proportional to the temperature gradient, and it would be reasonable now to consider that molecular mass transfer should be analogous to molecular energy transfer and that a corresponding relationship to Fourier's law of equation [2.3.1] should hold. For the molecular diffusion process this law is known as *Fick's law* and is that

$$J_{A_x} = -D_{AB} \frac{dc_A}{dx} \qquad\qquad [2.5.5]$$

Here J_{A_x} is the molar flux in the direction OX for a surface with normal in the OX direction, dc_A/dx is the gradient of c_A in the OX direction and D_{AB} is the *mass diffusivity* or *diffusion coefficient* for component A diffusing through component B. Substitution for J_A and c_A in equation [2.5.5] using equations [2.5.2] and [2.5.4] gives the corresponding result

$$j_{A_x} = -D_{AB} \frac{d\rho_A}{dx} \qquad\qquad [2.5.6]$$

Fick's law does not hold when there are significant temperature or pressure gradients within the fluid, since these in themselves can result in a mass diffusion. The other important point to note is that the diffusion coefficient D_{AB} is a property of the *pair* of components and can vary with the concentration, pressure and temperature. When A and B are gases, values for D_{AB} can vary between $5 \times 10^{-6} \, \text{m}^2/\text{sec}$ and $1.7 \times 10^{-4} \, \text{m}^2/\text{sec}$ for atmospheric temperature and pressure conditions, but for diffusion of a solute in a liquid such as water, values lie in the range $0.5 \times 10^{-9} \, \text{m}^2/\text{sec}$ to $3 \times 10^{-9} \, \text{m}^2/\text{sec}$.

2.6 Electric current flow

The flow of an electric current in many common materials is found to be in agreement with Ohm's law which, for a filament of uniform material, may be written

$$J = -\frac{1}{\rho} \frac{d\phi}{dx} \qquad [2.6.1]$$

where J is the current flow along the filament per unit cross-sectional area, $d\phi/dx$ is the gradient of the electric field and ρ is the resistivity. Many materials satisfy equation [2.6.1] and have a value for ρ which is sensibly independent of $d\phi/dx$, although it is in most cases dependent on the temperature of the material. Such materials are called *ohmic* and include the common conducting metals such as copper and also many insulating materials such as glass and plastic products. The ohmic materials exhibit a wide variation in resistivity, with typical extreme values being 1.51×10^{-8} ohm-m for silver, and 1×10^{16} ohm-m for quartz. Materials with low values, say below 1×10^{-6} ohm-m, are classed as good *conductors* and at the other extreme of the range we have the *insulating* materials.

The manner in which electric charge is transported in a solid conductor is related to the properties of the atom or molecule from which the material is made and also to the way in which these atoms or molecules are grouped together in the material. A solid material which is a good conductor has *free electrons* which move around from place to place and are not bound to a particular atom or molecule. In the presence of an electric field the free electrons can move and transport charge through the material under the influence of the electric field. Clearly, the rate of charge transport will depend upon the number of free electrons available to do this and also on their mobility and freedom within the atom or molecular structure. These factors vary from material to material and give rise to the wide variation in resistivity values. These factors are dependent on the internal energy of the material, and consequently give rise to the change in resistivity of a material with change in temperature.

Among the class of materials which do not satisfy equation [2.6.1], with ρ sensibly independent of the electric field gradient, are the *semiconductors*. In these materials we find that the resistivity depends upon $d\phi/dx$ and in many cases equation [2.6.1] has to be replaced by an equation of the type

$$J = -\alpha \left(\frac{d\phi}{dx}\right)^{\beta} \qquad [2.6.2]$$

where now α and β are constants, with β taking values between 1 and 5. Typical semiconductor materials are silicon, germanium and lead sulphide, all of which satisfy an equation similar to equation [2.6.2] and with β depending to a large extent on the method of manufacture of the material. The semiconductor materials cannot be said to have an inherent value of β for each material, since their electrical properties depend critically on the presence of minute quantities of contaminants.

When an electric current passes through any material it results in a dissipation of energy $\dot{\Phi}_e$ per unit volume given by

$$\dot{\Phi}_e = -J\frac{d\phi}{dx} \qquad [2.6.3]$$

For ohmic materials we may use equation [2.6.1] and obtain

$$\dot{\Phi}_e = \rho J^2 \qquad [2.6.4]$$

thus it can be seen that providing ρ is a positive quantity, then the dissipation of energy is necessarily positive. Normally the dissipation results in an increase in internal energy of the material and the temperature correspondingly rises.

Example 2.6.1
A copper wire of diameter 1 mm and length 6 m passes a current of 2·5 A. If the resistivity of copper is $1·56 \times 10^{-8}$ ohm-m, find the potential difference between the ends of the wire and the energy dissipation in the wire.

Data
$l = 6$ m

$d = 1$ mm

$I = 2·5$ A

$\rho = 1·56 \times 10^{-8}$ ohm-m

Analysis

$$J = \frac{I \times 4}{\pi d^2}$$

$$\Delta\phi = -l\frac{d\phi}{dx}$$

Therefore, substituting in equation [2.6.1],

$$\frac{I \times 4}{\pi d^2} = \frac{-\Delta\phi}{\rho l}$$

or

$$\Delta\phi = \frac{-4\rho l I}{\pi d^2}$$

$$\dot{\Phi}_e = \rho J^2 \times \frac{\pi d^2 l}{4}$$

$$= \frac{4\rho l I^2}{\pi d^2}$$

Calculation

$$\Delta\phi = \frac{4 \times 1 \cdot 56 \times 10^{-8} \times 6 \times 2 \cdot 5}{\pi \times 10^{-6}}$$

$$= \mathbf{0 \cdot 298 \ V}$$

$$\dot{\Phi}_e = \frac{4 \times 1 \cdot 56 \times 10^{-8} \times 6 \times 2 \cdot 5^2}{\pi \times 10^{-6}}$$

$$= \mathbf{0 \cdot 745 \ W}$$

2.7 Problems

1 A substance has a specific heat at constant volume that varies with the absolute temperature according to the relationship $c_v = \alpha + \beta T + \gamma T^2$. Find the change in internal energy when the substance is heated at constant volume from a temperature of 300 K to 425 K.

2 Using the data of Fig. 2.2.6 show that the specific heat at constant pressure for superheated steam does vary significantly with temperature.

3 A concrete wall of a building has a surface area of 120 m². The temperatures of the inner and outer surfaces are 293 K and 280 K on a typical day in winter. If the wall is 0·1 m in thickness and the thermal conductivity of the material is 1·1 W/m K find the heat transfer through the wall over a 24 hr period.

4 A door 2 m high and 0·7 m wide has a gap of 0·3 mm between it and the surrounding doorframe. The door is 50 mm thick and a flow of air passes through the narrow passage around the door. If the velocity of the air through the passage satisfies the equation

$$v = 6 \cdot 75x - 2 \cdot 25x^2$$

where x is the distance measured outward from the door find the following:

(a) the mass flow of air through the gap;
(b) the frictional dissipation per unit mass of air at the edge of the door and midway between the door and frame.

Assume the density and viscosity of the air are 1.225 kg/m^3 and 1.790 kg/m sec respectively.

5 A photographic studio has installed 1200 W of tungsten lighting. The studio gets hot and clients rarely return for a second session. The photographer takes advice and decides to remove the existing lighting and install a new type that gives improved illumination and is said to run cool. He installs 1200 W of lighting and is delighted with the improvement in his photography. Would his clients now return?

Chapter 3

The laws of thermodynamics and their application

3.1 Introduction

In the two previous chapters we have developed, through the conservation laws and property relationships, the fundamentals for a more detailed study of thermodynamic processes. We shall do this using what are known as the *laws of thermodynamics*. These laws, some of which we have already introduced in Chapter 1 but not named as such, determine the manner in which processes involving heat and work interactions may proceed. Other types of process are also involved in thermodynamics, but the main concern will be with heat and work processes.

Inevitably, laws of nature can be expounded in many different ways, each of which has its band of proponents, and no exception to this is found in the study of thermodynamics. Indeed, within thermodynamics there is perhaps more controversy over the fundamental laws than in any other developed branch of science. One of the reasons for this is that to some practitioners the study of the laws of thermodynamics is almost a theology and, as in a theology, personal opinions and beliefs tend to be held strongly and alternative opinions or approaches viewed with contempt. The attitude in this present discussion is that such passion is unhelpful in an introduction to the subject, and consequently the laws will be presented and discussed in a manner which is appropriate to the final objective – the application of the laws to practical processes.

3.2 Temperature and the zeroth law

The measurement of temperature is such a part of everyday life that a detailed discussion of it might at first seem quite unnecessary. Common methods of measurement include the use of thermometers, for example mercury or alcohol in glass, electrical resistance elements, bi-metal thermocouple junctions and a semiconductor element known as a thermistor. In each of these methods some observable or measurable characteristic of a substance – or construction of substances – is known to vary monotonically with hotness or coldness. The instrument is then calibrated at two reproducible temperatures which could be, for instance, the boiling and freezing points of a liquid at a fixed pressure. A linear scale can then be drawn between these two end points and the scale of temperature is established. Let us suppose that this has been done very accurately for each instrument device and that the fixed datum points are arbitrarily taken to be 21 and 107 units of temperature on this new scale. A bath of fluid at some intermediate hotness between 21 and 107 units is now used to check the reading of each instrument simultaneously. The results of such an experiment, however accurately carried out, might look like this:

Instrument type	A	B	C	D	E
Reading	73·1	74·0	72·9	73·2	73·0

The problem that now comes to the surface is quite simply, What is causing this disparity, and, more important, which one is correct? The cause of the problem is that in each case a linear scale has been arranged between the end points and in doing so it has been tacitly assumed that the various materials do behave in this linear manner as the temperature is increased. The confusing values obtained at an intermediate temperature simply serve to show that this is just not valid for most materials which do exhibit some form of measurable change with temperature. What is required is an *absolute scale of temperature*, i.e. a scale that is independent of all material properties. A later section will show how it is possible to obtain an absolute *thermodynamic* scale of temperature, but this can only be done when work on the second law of thermodynamics has been done. This analysis will also show that one example of such a scale is the perfect gas temperature scale. We shall use this scale for the moment and presume the later validation of this. In practice the perfect gas scale uses a thermometric device consisting of a closed container filled with gas at a pressure which ensures that, at the temperature range for which it is to be used, the gas is far removed from its liquefaction point. The product of the pressure and the volume of the gas is then proportional to the absolute temperature. All absolute scales, however devised, have a common zero point and it is only necessary to decide upon one datum point usually fixed by international agreement. The Kelvin scale of temperature is the usual one employed in the metric or SI system of units, and is the one we shall use here.

Given for the moment that such an absolute scale exists and can be proved in our subsequent work to be an absolute scale, we can now bring into the discussion the zeroth law of thermodynamics. It is so called because historically it was only realised to be a fundamental law of thermodynamics some time after the first law had been affirmed and so named. However, it is generally regarded as having a prior position to the first law and as a result is now listed as the zeroth law.

The zeroth law is concerned with a new notion or concept known as thermal equilibrium. Two bodies or systems are said to be in a state of *thermal equilibrium* if they are at the same *absolute temperature*. It is implied in this that if two bodies at the same temperature are brought together no heating or cooling takes place: this is the state of thermal equilibrium. In effect, this is a statement of the *zeroth law*. It may be deduced from this that if two bodies or systems are each separately in thermal equilibrium with a third, then all three must, in pairs or together, be in the equilibrium state. Each time we compare temperatures using a thermometric device we are effectively invoking the zeroth law of thermodynamics. In much of the analysis given in Chapter 2 we have been using this law without even stating it. That this has been possible without causing confusion is indicative of how deeply involved in everyday experience is this law. For other laws of thermodynamics the situation is perhaps a little different as will be shown in the subsequent sections.

3.3 First law, heat and work processes

The first law is concerned with energy and embodies the principle of conservation of energy. It acknowledges that various forms of energy exist in matter and that it is meaningful to talk of the total energy as being the sum of all the energy contributions. The principle of conservation of the total energy of a body of matter or a system, together with the notion of energy transfer by heat or work, is the second essential part of the first law of thermodynamics. Many of the equations derived in section 1.5, embody each of these aspects of the first law and *are* mathematical statements of the law. Thus for example equation [1.5.9], namely

$$\frac{d\mathcal{E}}{dt} = \dot{Q} - \dot{W} + \sum_{\text{in}} \dot{M}(u + pv + (\text{k.e.}) + \phi + \ldots) -$$

$$- \sum_{\text{out}} \dot{M}(u + pv + (\text{k.e.}) + \phi + \ldots)$$

is one possible way of expressing the first law of thermodynamics, as is the equation of thermal interaction, equation [1.5.15]

$$\Delta Q = \delta u + p \, \delta v - \Delta \Phi$$

The existence of many possible ways of expressing the law in mathematical form can cause confusion unless it is fully understood that the law is basically the principle of conservation of energy, and the mathematical expressions represent the application of the principle, or law, to particular systems. In the subsequent analyses the main concern will be to consider the application of the law in any of its mathematical forms to a number of practical systems. In this work it will be necessary to use much of the work in sections 2.2, 2.3 and 2.4 which, together with the first law, make extremely detailed analysis and investigation possible. Much of this work will be in the form of worked examples, which is helpful since practical systems can differ so widely that it is largely fruitless attempting to provide general equations for them. The following examples represent typical practical problems in which solutions can be obtained using a first-law analysis.

Example 3.3.1
An axial compressor has a mass flow rate of air of 50 kg/sec at inlet temperature and pressure of 288 K and 101 kN/m². The corresponding outlet conditions are 363 K and 180 kN/m². The outer surfaces of the compressor are exposed to the atmosphere and as a result there is a cooling amounting to a rate of 4 kW. Assuming little change in velocity between inlet and outlet find the power required to drive the compressor. In addition, if the rotational speed of the machine is 3000 r.p.m., find the torque in the drive shaft.

A schematic diagram of the compressor is given in the figure and shows air inlets and outlets at sections 1 and 2 respectively, together with the energy transfers $\dot{Q}$ and $\dot{W}$ by heating and working. Note that $\dot{Q}$ and $\dot{W}$ have arrowed directions for the energy transfer which accord with the convention outlined in section 1.5. In this particular problem $\dot{Q}$ will take a *negative* value since the actual direction of energy transfer by heating is *outward* from the compressor.

Data

$T_1 = 288$ K,	$T_2 = 363$ K
$p_1 = 101$ kN/m²,	$p_2 = 180$ kN/m²
$\dot{Q} = -4$ kW,	$\dot{M} = 50$ kg/sec
$v_1 = v_2$,	$c_p = 1 \cdot 0$ kJ/kg K

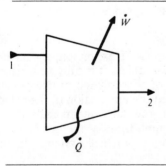

Analysis

Applying the steady-flow energy equation [1.5.9] to the system we have

$$\frac{d\mathscr{E}}{dt} = \dot{Q} - \dot{W} + \dot{M}_1 \left(u_1 + p_1 v_1 + \frac{v_1^2}{2} \right) - \dot{M}_2 \left(u_2 + p_2 v_2 + \frac{v_2^2}{2} \right)$$

but

$$\frac{d\mathscr{E}}{dt} = 0, \qquad \dot{M}_1 = \dot{M}_2, \qquad v_1 = v_2 \quad \text{and} \quad (u + pv) = h = c_p T$$

Therefore

$$\dot{W} = \dot{Q} + \dot{M}c_p(T_1 - T_2)$$

Note that $\dot{W}$ does not depend on either p_1 or p_2.

Calculation

$$\dot{W} = -4 + 50 \times 1{\cdot}0 \times (288 - 363)$$

$$= -4 - 3750$$

$$= \mathbf{-3754\ kW}$$

The negative sign indicates a power input to the compressor, as would be expected.

For the torque τ in kNm we have

$$-\dot{W} = \frac{2\pi\tau}{60} \times (\text{r.p.m.})$$

or

$$\tau = \frac{60 \times 3754}{2\pi \times 3000}$$

$$= \mathbf{1{\cdot}195\ kNm}$$

Example 3.3.2

A jet engine in an aircraft operates at the following conditions. At inlet the temperature is 220 K, the velocity 120 m/sec, and at outlet the temperature is 900 K, and the velocity 400 m/sec. The heat input to the combustion chamber is 40 MW and the total power offtake from the engine for aircraft supplies is 100 kW. Find the engine mass flow using this data, assuming an average value for c_p of $1{\cdot}0$ kJ/kg K and neglecting any increase in mass flow due to the fuel supply.

Data

$T_1 = 220$ K, $T_2 = 900$ K

$u_1 = 120$ m/sec, $u_2 = 400$ m/sec

$\dot{Q} = 40$ MW, $\dot{W} = 100$ kW

$\dot{M}_1 = \dot{M}_2$, $c_p = 1{\cdot}0$ kJ/kg K

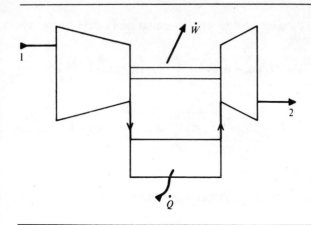

Analysis

The steady-flow energy equation [1.5.9] applied to the system with $d\mathscr{E}/dt = 0$ and $h = (u + pv)$ gives

$$0 = \dot{Q} - \dot{W} + \dot{M}\left\{(h_1 - h_2) + \left(\frac{v_1^2}{2} - \frac{v_2^2}{2}\right)\right\}$$

$$= \dot{Q} - \dot{W} + \dot{M}\left\{c_p(T_1 - T_2) + \left(\frac{v_1^2}{2} - \frac{v_2^2}{2}\right)\right\}$$

Calculation

$$0 = 40\,000 - 100 + \dot{M}\left\{1{\cdot}0(220 - 900) + \right.$$

$$\left. + \left(\frac{120^2}{2} - \frac{400^2}{2}\right) \times 10^{-3}\right\} \text{ kW}$$

Note: The 10^{-3} multiplier in the kinetic energy term is necessary for the kinetic energy to be in kJ/kg rather than J/kg.

$$0 = 39\,900 - \dot{M}(680 + 72{\cdot}8)$$

$$\dot{M} = \textbf{53}{\cdot}\textbf{0 kg/sec}$$

Example 3.3.3

A heat exchanger, thermally insulated on its outer surface, is designed to cause heat transfer between a stream of air and a stream of carbon dioxide. The two gases are separated in the device by either metal plates or tube walls. The required inlet and outlet conditions of the two gases are given in the

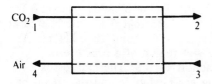

data below and the problem is to find the mass flow ratio of the two gases that is required to produce these conditions.

Data

$T_1 = 75\,°C$, $T_2 = 50\,°C$

$T_3 = 15\,°C$, $T_4 = 60\,°C$

v_1, v_2, v_3, v_4 all small

$\dot{Q} = \dot{W} = 0$

$c_{p(air)} = 1·0$ kJ/kg K, $c_{p(CO_2)} = 0.85$ kJ/kg K

Analysis

The steady-flow energy equation [1.5.9] gives for the system, with $h = (u + pv)$,

$$0 = \dot{M}_1 h_1 - \dot{M}_2 h_2 + \dot{M}_3 h_3 - \dot{M}_4 h_4$$

but from mass conservation principles

$$\dot{M}_1 = \dot{M}_2, \qquad \dot{M}_3 = \dot{M}_1$$

Therefore

$$\dot{M}_1 (h_1 - h_2) = \dot{M}_3 (h_4 - h_3)$$

or

$$\dot{M}_1 c_{p(CO_2)} (T_1 - T_2) = \dot{M}_3 c_{p(air)} (T_4 - T_3)$$

Note: The numerical value of $(T_1 - T_2)$ is the same regardless of whether the temperature unit is in Celsius or on the absolute Kelvin scale.

Calculation

$$\dot{M}_1 \times 0·85 (75 - 50) = \dot{M}_3 \times 1·0 (60 - 15)$$

$$M_1/M_3 = 2·118$$

The required mass flow ratio is therefore

$$\frac{\text{mass flow rate } CO_2}{\text{mass flow rate air}} = \mathbf{2·118}$$

Example 3.3.4

A steam turbine has a mass flow rate of 400 kg/sec and the inlet condition of the steam is such that it is superheated at a temperature and pressure of 400°C and 5000 kN/m². At outlet the steam is in the wet region at a pressure of 20 kN/m² and dryness fraction x = 0·9. Thermal losses through the turbine casing are negligible and the steam has approximately the same inlet and outlet velocity. Find the power output of the turbine.

Data

$$p_1 = 5000 \text{ kN/m}^2, \qquad T_1 = 400 \text{ °C}$$

$$p_2 = 20 \text{ kN/m}^2, \qquad x_2 = 0\cdot9$$

$$\dot{Q} = 0, \qquad\qquad \dot{M} = 400 \text{ kg/sec}$$

$$v_1 = v_2$$

Analysis

Energy balance

$$0 = \dot{Q} - \dot{W} + \dot{M}(h_1 - h_2)$$

but $\dot{Q} = 0$, therefore

$$\dot{W} = \dot{M}(h_1 - h_2)$$

The value of h_1 is found directly from tables of steam properties in the superheat region, h_2 is found using the relationship of equation [2.2.26], i.e.

$$h_2 = (1 - x_2)h_{12} + x_2 h_{g2}$$

Values of h_{12} and h_{g2} are found from tables of steam properties in the wet region.

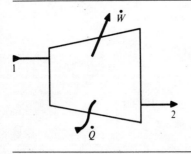

Calculation

From tables

$h_1 = 3196 \text{ kJ/kg}$

$h_{12} = 251 \text{ kJ/kg}, \qquad h_{g2} = 2609 \text{ kJ/kg}$

Therefore

$h_2 = 0 \cdot 1 \times 251 + 0 \cdot 9 \times 2609$

$\quad = 2373 \text{ kJ/kg}$

$\dot{W} = 400(3196 - 2373) \text{ kW}$

$\quad = \textbf{329} \cdot \textbf{2 MW}$

The above examples illustrate the application of the steady-flow energy equation to a number of practical thermodynamic systems. An important point to note is that the steady-flow energy equation is valid quite independently of whether there is large or negligible frictional dissipation within a system. Each answer is numerically exact providing only that the input data are correct.

It is interesting now to consider a steady-flow system where the frictional dissipation may be substantial and indeed is deliberately so.

Steam throttle

In its simplest form a steam throttle consists of a steam pipe in which some obstruction is placed such as an orifice plate, wire mesh or even a half-shut valve. Steam velocities are normally little different on either side of the obstruction and if suffixes 1 and 2 are used for inlet and outlet conditions then the energy equation

$$0 = \dot{Q} - \dot{W} + \dot{M}\left(h_1 + \frac{v_1^2}{2}\right) - \dot{M}\left(h_2 + \frac{v_2^2}{2}\right)$$

reduces to the simple form

$h_1 = h_2$

However, the effect of the frictional dissipation is to cause a pressure drop in the direction of flow and thus $p_1 > p_2$. Consider now the implication of this if the inlet steam flow is at pressure $p_1 = 500 \text{ kN/m}^2$, dryness fraction x = 0·98, and at outlet the pressure is reduced to 100 kN/m^2. We have

$h_{11} = 640 \text{ kJ/kg}, \qquad h_{g1} = 2749 \text{ kJ/kg}$

Therefore

$$h_1 = 0.02 \times 640 + 0.98 \times 2749$$
$$= 2706 \text{ kJ/kg}$$

thus

$$h_2 = h_1 = 2706 \text{ kJ/kg} \quad \text{at} \quad p_2 = 100 \text{ kN/m}^2.$$

Examination of the tables of properties of steam shows that at this pressure a value of $h_2 = 2706$ kJ/kg cannot be achieved within the wet steam region since at $p_2 = 100$ kN/m^2, $h_{g2} = 2675$ kJ/kg $< h_2$. The steam, after flowing through the throttle, is therefore *superheated* but at the same enthalpy as before the throttle, where it was in the wet region.

The steam-throttling process provides a convenient and often-used method for providing superheated steam where this is required, and perhaps, where the boiler supplying the steam supply does not incorporate a superheater to do this.

In cases where the steam supply has a low dryness fraction it may prove impossible to use the steam throttle by itself to produce superheated steam, or the pressure drop required may be quite unacceptable. In such cases the solution is to use some form of separator before the throttle to remove some of the liquid prior to throttling. There are many designs for separating plant, most of which rely on some form of centrifugal action such as passing the steam flow round a sharp bend and collecting the water ejected in droplet form from the main flow.

The piston and cylinder

So far the examples in this section have all been concerned with systems in which there was a flow through the device or system. A common system which now needs to be considered is the simple piston and cylinder shown in Fig. 3.3.1. It is assumed that the cylinder may be being heated and that during a small movement of

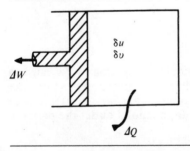

Fig. 3.3.1. Energy transfer in the piston and cylinder.

the piston the changes in kinetic and potential energy of the piston and potential energy changes of the material in the cylinder can be neglected. Using the notation in the figure the energy balance is simply

$$0 = \Delta Q - \Delta W - \delta u - \delta\left(\frac{v^2}{2}\right) \qquad [3.3.1]$$

where δu and ΔW are the change in internal energy and work done per unit mass.

However, we have already shown in section 1.5 that when a small mass is introduced across a system boundary, the element of work done per unit mass is ΔW and a brief study of Fig. 3.3.1 shows the situation to be analogous. The only difference here is that there may be significant frictional dissipation between the piston and the cylinder walls, and if this is of amount $\Delta\Phi$ per unit mass,

$$\Delta W = p\,\delta v - \Delta\Phi \qquad [3.3.2]$$

Note that this equation is valid regardless of whether the process is that of expansion or compression, providing $\Delta\Phi$ is always taken to be a positive energy dissipation. Combination of equations [3.3.1] and [3.3.2] gives finally

$$\Delta\dot{Q} = \delta u + p\,\delta v + \delta\left(\frac{v^2}{2}\right) - \Delta\Phi \qquad [3.3.3]$$

In many instances it will be possible to neglect changes in kinetic energy of the material within the cylinder, in which case one obtains the simpler expression

$$\Delta Q = \delta u + p\,\delta v - \Delta\Phi \qquad [3.3.4]$$

which together with equation [3.3.2] is the basis on which problems concerned with piston and cylinder devices can be dealt with.

Comparison of equation [3.3.4] with the equation of thermal interaction [1.5.15] shows that they appear to be identical. Note, however, that the general result equation [3.3.3] is *not* identical with the equation of thermal interaction.

Example 3.3.5
Air initially at a temperature of 300 K is compressed slowly in a piston and cylinder from 100 kN/m^2 to 500 kN/m^2 pressure. During the compression the pressure and specific volume are approximately related by $pv^{1.3} = \text{constant} = k$, and 20% of the work done on the piston is accounted for by frictional dissipation at the cylinder walls. Find the total work done in the process and the corresponding heating or cooling.

Data

$$p_1 = 100 \text{ kN/m}^2, \qquad p_2 = 500 \text{ kN/m}^2$$

$$pv^{1 \cdot 3} = k$$

$$\int_1^2 \Delta\Phi = -0 \cdot 2 \int_1^2 \Delta W$$

$$c_v = 0 \cdot 718 \text{ kJ/kg K}, \qquad R = 0 \cdot 287 \text{ kJ/kg K}$$

Analysis

Equation [3.3.2] gives for elemental process

$$\Delta W = p \, \delta v - \Delta\Phi$$

or

$$\int_1^2 \Delta W = \int_1^2 p \, dv - \int_1^2 \Delta\Phi$$

Therefore

$$\int_1^2 \Delta W + \int_1^2 \Delta\Phi = \int_1^2 p \, dv$$

thus

$$0 \cdot 8 \int_1^2 \Delta W = \int_1^2 p \, dv = k \int_1^2 v^{-1 \cdot 3} \, dv = \frac{(p_1 v_1^{1 \cdot 3})}{0 \cdot 3} \left\{ \frac{1}{v_1^{0 \cdot 3}} - \frac{1}{v_2^{0 \cdot 3}} \right\}$$

Equation [3.3.4] gives

$$\Delta Q = \delta u + p \, \delta v - \Delta\Phi = \delta u + \Delta W$$

or

$$\int_1^2 \Delta Q = (u_2 - u_1) + \int_1^2 \Delta W = c_v (T_2 - T_1) + \int_1^2 \Delta W$$

$$= c_v \left(\frac{p_2 v_2}{R} - T_1 \right) + \int_1^2 \Delta W$$

However, $p_1 v_1^{1 \cdot 3} = p_2 v_2^{1 \cdot 3}$, therefore

$$v_2 = v_1 \left(\frac{p_1}{p_2} \right)^{1/1 \cdot 3} = \frac{R T_1}{p_1} \left(\frac{p_1}{p_2} \right)^{1/1 \cdot 3}$$

Calculation

$$v_2 = \frac{0 \cdot 287 \times 300}{100} \left(\frac{100}{500}\right)^{0 \cdot 769} = 0 \cdot 250 \ \text{m}^3/\text{kg}$$

$$v_1 = \frac{0 \cdot 287 \times 300}{100} = 0 \cdot 861 \ \text{m}^3/\text{kg}$$

$$\int_1^2 \Delta W = \frac{100 \times 0 \cdot 861^{1 \cdot 3}}{0 \cdot 8 \times 0 \cdot 3} \left\{\frac{1}{0 \cdot 861^{0 \cdot 3}} - \frac{1}{0 \cdot 250^{0 \cdot 3}}\right\}$$

$$= -161 \ \text{kJ/kg}$$

$$\int_1^2 \Delta Q = 0 \cdot 718 \left(\frac{500 \times 0 \cdot 250}{0 \cdot 287} - 300\right) - 161$$

$$= -64 \cdot 3 \ \text{kJ/kg}$$

hence the cylinder is being cooled.

Real processes in a perfect gas

A convenience which is often used in thermodynamics, and which was introduced in Example 3.3.5, is that of approximating the relationship between pressure and specific volume during the process by an equation of the form $pv^n = k$, where n is the index of expansion or compression. It is necessarily an approximation, but despite this can prove extremely useful and indeed is in widespread use. Previously in section 2.2 we showed that for an adiabatic frictionless process of a perfect gas, i.e. $\Delta Q = 0$, $\Delta \Phi = 0$ the pressure and specific volume do satisfy the equation $pv^n = k$ with $n = \gamma = (c_p/c_v)$ the specific heat ratio. It is interesting now to look into real processes of a perfect gas where ΔQ and $\Delta \Phi$ are not necessarily zero with the object of finding what magnitude the constant n might have, compared to γ.

We consider first the case where $\Delta \Phi \neq 0$ and $\Delta Q \neq 0$ for which equation [1.5.15] may be written

$$\Delta Q + \Delta \Phi = \delta u + p \, \delta v$$

or

$$\Delta Q + \Delta \Phi = c_v \, \delta T + p \, \delta v = \frac{c_v}{R} (p \, \delta v + v \, \delta p) + p \, \delta v \qquad [3.3.5]$$

and we have also

$$pv^n = k$$

or

$$npv^{n-1} \, \delta v + v^n \, \delta p = 0$$

or

$$np \, \delta v + v \, \delta p = 0 \qquad\qquad [3.3.6]$$

Combining equations [3.3.5] and [3.3.6] to eliminate the term in $v \, \delta p$ we obtain

$$\Delta Q + \Delta \Phi = p \, \delta v \left(\frac{c_v}{R} + 1 - \frac{nc_v}{R} \right)$$

$$= \frac{p \, \delta v}{R} (c_p - nc_v)$$

i.e.

$$\Delta Q + \Delta \Phi = \frac{c_v}{R} (\gamma - n) p \, \delta v \qquad\qquad [3.3.7]$$

Suppose now the process is an expansion process, i.e. $\delta v > 0$, and that the frictional dissipation $\Delta \Phi > 0$. If each side of equation [3.3.7] is to have the same sign, the following inequalities must be satisfied.

For $\Delta Q = 0$, $\Delta \Phi > 0$

then $\gamma > n$ (adiabatic process with friction)

$\Delta Q > 0$, $\Delta \Phi > 0$

then $\gamma > n$ (heating process with friction)

$0 > \Delta Q \geq -\Delta \Phi$, $\Delta \Phi > 0$

then $\gamma \geq n$ (cooling process with $\Delta Q \geq -\Delta \Phi$)

$\Delta Q < -\Delta \Phi$, $\Delta \Phi > 0$

then $\gamma < n$ (cooling process with $\Delta Q < -\Delta \Phi$)

For a compression process, i.e. $\delta v < 0$, the corresponding relationships are:

For $\Delta Q = 0$, $\Delta \Phi > 0$

then $\gamma < n$ (adiabatic process with friction)

$\Delta Q > 0$, $\Delta \Phi > 0$

then $\gamma < n$ (heating process with friction)

$0 > \Delta Q \geq -\Delta \Phi$, $\Delta \Phi > 0$

then $\gamma \leq n$ (cooling process with $\Delta Q \geq -\Delta \Phi$)

$\Delta Q < -\Delta \Phi$, $\Delta \Phi > 0$

then $\gamma > n$ (cooling process with $\Delta Q < -\Delta \Phi$)

Although the inequalities given above hold only strictly for a perfect gas they can be used for other media, particularly vapours, with reasonable accuracy if required. A suitable value for γ is required, and is arrived at *not* by considering specific heats but by taking the equation of thermal interaction with $\Delta Q = \Delta \Phi = 0$ and using it in combination with the approximation $pv^{\gamma} = $ constant to give a value for γ. Thus, we have

$$\delta u + p\, \delta v = 0$$

and from equation [3.3.6] with $n = \gamma$,

$$\gamma p\, \delta v + v\, \delta p = 0$$

and eliminating $p\, \delta v$ between the two equations the result is

$$\gamma \delta u - v\, \delta p = 0$$

Since for $\Delta Q = \Delta \Phi = 0$ the entropy S is constant we have

$$\gamma = v \left(\frac{\partial p}{\partial u} \right)_{s = \text{constant}} \qquad [3.3.8]$$

It is now not difficult, using tabulated values of p and u for a particular vapour, to determine the appropriate value for γ, the index for an adiabatic frictionless process.

A process for which the approximation $pv^{n} = $ constant *is* reasonably valid is called a *polytropic* process, and a number of relationships can be developed for $\int_1^2 \Delta W$ and $\int_1^2 \Delta Q$ for such a process using the equation of thermal interaction [1.5.15] and [3.3.2]. The type of analysis used is similar to that in Example 3.3.5 and will not be repeated. Two of the more useful equations resulting from this do not in any way depend on an assumption that the medium is a perfect gas and are

$$\int_1^2 \Delta W = \frac{-1}{(n-1)} (p_2 v_2 - p_1 v_1) - \int_1^2 \Delta \Phi \qquad [3.3.9]$$

$$\int_1^2 \Delta Q = (u_2 - u_1) + \int_1^2 \Delta W \qquad [3.3.10]$$

For a perfect gas it is easy to prove that the change in internal energy $(u_2 - u_1)$ is given by

$$u_2 - u_1 = \frac{1}{(\gamma - 1)} (p_2 v_2 - p_1 v_1) \qquad [3.3.11]$$

which together with equations [3.3.9] and [3.3.10] gives particularly convenient expressions for $\int_1^2 \Delta W$ and $\int_1^2 \Delta Q$. These must only be used when the perfect gas assumption is justified.

Adiabatic and frictionless flow of a perfect gas

There are many fluid flows where the effects of heat transfer and frictional dissipation are not of paramount importance. Examples include some high-velocity gas flows in turbo-machinery, and high Mach number flows over aircraft wings and fuselages. Frictional effects near the boundary surfaces of these flows *are* significant, and the attention in this work here is primarily directed at flow regions away from such surfaces.

We begin by considering steady flow along a stream tube having a constant area along its length and assume the flow process is frictionless. Since the flow area is constant the principle of conservation of mass gives

$$\rho v = \text{constant}$$

or

$$\rho \delta v + v \delta \rho = 0$$

Momentum conservation for a frictionless process with no work input or output gives, from equation [1.4.15],

$$v \delta v + \frac{\delta p}{\rho} = 0$$

Eliminating δv between the two results and rearranging, produces the interesting result that

$$v^2 = \frac{\delta p}{\delta \rho} \tag{3.3.12}$$

In other words the steady flow can only exist at the velocity determined by equation [3.3.12]. The flow represented here is one in which there are variations in pressure and density. It is steady and in fact represents the flow an observer would see if he were moving with a sound wave at the velocity of propagation of the sound wave, and observing the flow moving past him. The velocity

$$a = \left(\frac{\partial p}{\partial \rho} \right)^{1/2}_{\Delta \Phi = 0}$$

is therefore the velocity of propagation of a sound wave by a frictionless process in any material. The equation of thermal interaction for a frictionless process is from equation [1.5.15]

$$\Delta Q = \delta u + p \, \delta v$$

or

$$\Delta Q = \delta u - \frac{p \, \delta \rho}{\rho^2} \tag{3.3.13}$$

and thus the velocity of sound propagation a may be deduced as given by

$$a^2 = \frac{\delta p}{\delta \rho} = \frac{p \, \delta p}{\rho^2 (\delta u - \Delta Q)} \qquad [3.3.14]$$

The velocity of propagation thus depends on the heating or cooling of the medium through the term ΔQ in equation [3.3.14]. Little more can be done with equation [3.3.14] in the general case without appealing to tabulated values of properties, but in the case of a perfect gas further analysis is possible. We shall not look into the general case further, but consider the simplified problem where the medium is a perfect gas and where $\Delta Q = 0$. For this situation, then, equation [3.3.14] reduces to

$$a^2 = \frac{p}{\rho^2} \frac{\delta p}{\delta u} = \frac{p}{\rho^2} \frac{\delta p}{c_v \, \delta T}$$

which, together with equation [2.2.19] for an adiabatic frictionless process in a perfect gas, gives the final equation

$$a^2 = \frac{p}{c_v \rho^2} \cdot \frac{\gamma}{(\gamma - 1)} \cdot \frac{p}{T}$$
$$= \gamma R T \qquad [3.3.15]$$

which enables the value of the accoustic velocity a to be found in terms of the temperature and material properties R and γ. However, it is important to note that – as equation [3.3.14] clearly shows – the speed of propagation depends critically on heating or cooling and the value that can be calculated using equation [3.3.15] is for $\Delta Q = 0$. Care must also be taken to ensure that when inserting numerical values in equation [3.3.15] that both sides are in the same units. Thus if a is required in m/sec then the gas constant R must be in J/kg K and *not* kJ/kg K as has been used previously. Thus the speed of sound in air at temperature 300 K is, with $R = 0 \cdot 2871$ kJ/kg K,

$$a = (1 \cdot 4 \times 0 \cdot 2871 \times 10^3 \times 300)^{1/2}$$

$$= 347 \cdot 2 \text{ m/sec}$$

Much of the work necessary in dealing with high-velocity flow of gases or vapours is eased by the adoption of new parameters called stagnation properties, in particular stagnation pressure and temperature. We consider firstly the adiabatic flow of a perfect gas along a stream tube, for which the steady-flow energy equation reduces to

$$h + \frac{v^2}{2} = \text{constant}$$

or, say,

$$h + \frac{v^2}{2} = h_s \qquad [3.3.16]$$

where h_s as defined by equation [3.3.16] is called the *stagnation enthalpy* of a flow along the stream tube. The reasons for this are clear if one considers a subsequent section of the stream tube where the velocity is small or zero. Let the corresponding enthalpy be h', then from equation [3.3.16] we have

$$h' = h_s$$

and thus h_s, the stagnation enthalpy, is actually the enthalpy the fluid would have if it were brought to rest by an *adiabatic* process. Note it is not necessary to assume anything about the frictional dissipation in the flow and the result is quite independent of this. The definition of *stagnation temperature*, T_s, follows immediately by substituting for h and h_s in equation [3.3.16] to give

$$c_p T + \frac{v^2}{2} = c_p T_s$$

or

$$T_s = T + \frac{v^2}{2c_p} \qquad [3.3.17]$$

Again, the stagnation temperature is the temperature achieved by the fluid if it is brought to rest by an *adiabatic*, but not necessarily frictionless, process. On the other hand the other new parameter, *stagnation pressure*, can only be defined in terms of an adiabatic *and* frictionless process, and is that pressure achieved by the fluid when it is brought to rest by such a process. A process where both ΔQ and $\Delta \Phi$ are zero is, from equation [2.2.22], one in which the entropy remains constant and is called *isentropic*.

The relationship between pressure and temperature for an isentropic process in a perfect gas has been obtained previously (equation [2.2.19]) and is that

$$pT^{-\gamma/(\gamma-1)} = \text{constant}$$

Consequently, for a perfect gas the stagnation pressure, p_s, is given by

$$p_s T_s^{-\gamma/(\gamma-1)} = pT^{-\gamma/(\gamma-1)}$$

or

$$p_s = p\left(\frac{T_s}{T}\right)^{\gamma/(\gamma-1)} \qquad [3.3.18]$$

If we now define the Mach number M_n to be the ratio of the fluid velocity v to the local acoustic velocity a then we have

$$M_n = \frac{v}{a}$$

$$= \frac{v}{(\gamma RT)^{1/2}} \qquad [3.3.19]$$

In terms of the Mach number, equations [3.3.17] and [3.3.18] are found to have an interesting form. In the case of T_s we have, from equation [3.3.17], that

$$\frac{T_s}{T} = \left(1 + \frac{v^2}{2c_pT}\right) = \left(1 + \frac{\gamma RM_n^2}{2c_p}\right)$$

$$= \left(1 + \frac{(\gamma - 1)}{2} M_n^2\right) \qquad [3.3.20]$$

and correspondingly from equation [3.3.18]

$$\frac{p_s}{p} = \left(1 + \frac{(\gamma - 1)}{2} M_n^2\right)^{\gamma/(\gamma - 1)} \qquad [3.3.21]$$

Example 3.3.6
A jetliner is flying at $M_n = 0.9$ at an altitude of 10 000 m where the air temperature and pressure are 223·3 K and 26·5 kN/m² respectively. Find the corresponding temperature and pressure on the leading edge of the wing where the air velocity relative to the wing is negligible.

Data
$z = 10\ 000$ m
$M_n = 0.9$
$T = 223.3$ K
$p = 26.5$ kN/m²
$\gamma = 1.4$

Analysis
Assume adiabatic and frictionless flow to the wing leading edge and thus the resulting temperature and pressure are the stagnation values. We have then

$$T_s = T\left(1 + \frac{(\gamma - 1)}{2}M_n^2\right)$$

and

$$p_s = p\left(1 + \frac{(\gamma - 1)}{2} M_n^2\right)^{\gamma/(\gamma - 1)}$$

Calculation

$$T_s = 223 \cdot 3 \left(1 + \frac{(1 \cdot 4 - 1)}{2} \times 0 \cdot 9^2 \right)$$

$$= \mathbf{259 \cdot 5 \ K}$$

$$p_s = 26 \cdot 5 \left(1 + \frac{(1 \cdot 4 - 1)}{2} \times 0 \cdot 9^2 \right)^{1 \cdot 4/0 \cdot 4}$$

$$= \mathbf{44 \cdot 82 \ kN/m^2}$$

Consider now the flow of a perfect gas at velocity v normal to a cross-section of area A. The mass flow rate through the cross-section is $\dot{M}$ given by equation [1.2.5], i.e.

$$\dot{M} = \rho A v$$

or

$$\dot{M} = \rho A M_n a = \frac{p A M_n}{RT} (\gamma R T)^{1/2}$$

but

$$p_s = p \left(1 + \frac{(\gamma - 1)}{2} M_n^2 \right)^{\gamma/(\gamma - 1)} \quad \text{and} \quad T_s = T \left(1 + \frac{(\gamma - 1)}{2} M_n^2 \right)$$

and thus

$$\dot{M} = \left(\frac{\gamma}{R}\right)^{1/2} \cdot \left(\frac{p_s A M_n}{T_s^{1/2}}\right) \cdot \left(1 + \frac{(\gamma - 1)}{2} M_n^2 \right)^{-(\gamma + 1)/2(\gamma - 1)}$$

or

$$\frac{\dot{M} T_s^{1/2}}{A p_s} = \left(\frac{\gamma}{R}\right)^{1/2} M_n \left(1 + \frac{(\gamma - 1)}{2} M_n^2 \right)^{-(\gamma + 1)/2(\gamma - 1)} \qquad [3.3.22]$$

For given values of A, p_s and T_s the mass flow rate will be a maximum or minimum when

$$\frac{d}{dM_n} \left\{ M_n \left(1 + \frac{(\gamma - 1)}{2} M_n^2 \right)^{-(\gamma + 1)/2(\gamma - 1)} \right\} = 0$$

which occurs when $M_n = 1 \cdot 0$ and can be shown to be a *maximum* value. The corresponding value of $MT_s^{1/2}/Ap_s$ is called the *critical value* and is given by

$$\left(\frac{\dot{M} T_s^{1/2}}{A p_s}\right)_{\text{critical}} = \left(\frac{\gamma}{R}\right)^{1/2} \cdot \left\{1 + \frac{(\gamma - 1)}{2} \right\}^{-(\gamma + 1)/2(\gamma - 1)} \qquad [3.3.23]$$

The *critical pressure* ratio is similarly defined by substituting $M_n = 1 \cdot 0$ in equation [3.3.21] and is

$$\left(\frac{p_s}{p}\right)_{\text{critical}} = \left(1 + \frac{(\gamma - 1)}{2}\right)^{\gamma/(\gamma - 1)} \tag{3.3.24}$$

The phenomena whereby the mass flow rate through an area has a maximum value at a flow Mach number of $1 \cdot 0$ is called *choking* and, if the area is a converging nozzle, then at this condition the nozzle is said to be *choked*. The pressure ratio (p_s/p) in the exit plane of a convergent nozzle can never exceed the critical value, and consequently if the nozzle discharges into an ambient pressure p_a such that $(p_s/p_a) > (p_s/p)_{\text{critical}}$ then some of the expansion to p_a must take place after the fluid has left the nozzle exit.

Example 3.3.7
Air flows through a convergent nozzle at a stagnation pressure and temperature of $300 \, \text{kN/m}^2$ and $400 \, \text{K}$ respectively. The exit area of the nozzle is $0 \cdot 1 \, \text{m}^2$. Find the mass flow rate, exit temperature, pressure and Mach number for discharge into an ambient pressure of (a) $200 \, \text{kN/m}^2$, (b) $100 \, \text{kN/m}^2$.

Data
$p_s = 300 \, \text{kN/m}^2$, $T_s = 400 \, \text{K}$
$A = 0 \cdot 1 \, \text{m}^2$
(a) $p_a = 200 \, \text{kN/m}^2$
(b) $p_a = 100 \, \text{kN/m}^2$

Analysis
$$\left(\frac{p_s}{p}\right)_{\text{critical}} = \left\{1 + \frac{(\gamma - 1)}{2}\right\}^{\gamma/(\gamma - 1)}$$

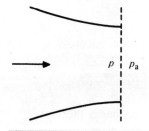

$$p \quad p_a$$

Check to see in each case whether $(p_s/p_a) > (p_s/p)_{critical}$. If not then $p = p_a$ and we have from equation [3.3.21] and [3.3.20] that

$$\frac{p_s}{p_a} = \left(1 + \frac{(\gamma - 1)}{2} M_n^2\right)^{\gamma/(\gamma - 1)}$$

which gives a value for M_n.

$$\frac{T_s}{T} = \left\{1 + \frac{(\gamma - 1)}{2} M_n^2\right\} = \left(\frac{p_s}{p}\right)^{(\gamma - 1)/\gamma}$$

giving the exit temperature T.

Finally, we have from equation [3.3.22] that

$$\dot{M} = \frac{Ap_s}{T_s^{1/2}} \left(\frac{\gamma}{R}\right)^{1/2} M_n \left\{1 + \frac{(\gamma - 1)}{2} M_n^2\right\}^{-(\gamma + 1)/2(\gamma - 1)}$$

from which $\dot{M}$ can be calculated.

If $(p_s/p_a) > (p_s/p)_{critical}$ then the nozzle is choked, the pressure in the exit plane is the critical value and not p_a, the exit Mach number is unity and this value may then be used as above in equations [3.3.20] and [3.3.22] to find T and $\dot{M}$.

Calculation

$$\left(\frac{p_s}{p}\right)_{critical} = \left\{1 + \frac{(1 \cdot 4 - 1)}{2}\right\}^{1 \cdot 4/0 \cdot 4} = 1 \cdot 893$$

For (a):

$$\left(\frac{p_s}{p_a}\right) = \frac{300}{200} = 1 \cdot 5 < \left(\frac{p_s}{p}\right)_{critical} \qquad \text{therefore} \qquad \boldsymbol{p_{exit} = p_a}$$

Thus

$$1 \cdot 5 = \left\{1 + \frac{(1 \cdot 4 - 1)}{2} M_n^2\right\}^{3 \cdot 5}$$

Hence

$$M_n = \boldsymbol{0 \cdot 784}$$

$$I_s = 1 \cdot 5^{1/3 \cdot 5} = 1 \cdot 12$$

Therefore

$$T = \boldsymbol{357\ K}$$

$$\dot{M} = \frac{0 \cdot 1 \times 300}{400^{1/2}} \left(\frac{1 \cdot 4 \times 10^3}{0 \cdot 287}\right)^{1/2} \times 0.784 \left(1 + \frac{0 \cdot 4}{2} \times 0.784^2\right)^{-2 \cdot 4/0 \cdot 8}$$

Note: R is in kJ/kg K and the 10^3 multiplier is therefore required.

Therefore

$\dot{M} = \mathbf{57\cdot9\ kg/sec}$

For (b):

$$\left(\frac{p_s}{p_a}\right) = \frac{300}{100} = 3 > \left(\frac{p_s}{p}\right)_{\text{critical}}$$

hence the nozzle is choked and $p = 300/1\cdot893$. Therefore

$p = \mathbf{158\ kN/m^2}$

$M_n = \mathbf{1\cdot0}$

and

$$T = \frac{T_s}{\left(1 + \dfrac{0\cdot4}{2}\right)} = \frac{400}{1\cdot2}$$

Therefore

$T = \mathbf{333\ K}$

$$\dot{M} = \frac{0\cdot1 \times 300}{400^{1/2}} \left(\frac{1\cdot4 \times 10^3}{0\cdot287}\right)^{1/2} \times 1\cdot0 \times 1\cdot2^{-3}$$

$\qquad = \mathbf{60\cdot6\ kg/sec}$

3.4 Second law and the Carnot theorem

In many previous sections we have been concerned with laws of nature, all of which could be stated mathematically in the form of simple equations involving equalities. Generally, the laws concerned with conservation principles resulted in exact mathematical relationships and the constitutive laws of materials gave rise to realistic approximations of actual physical processes. The second law of thermodynamics is quite unlike any of these previous relationships since it is concerned with the *direction* in which processes take place as time increases. Although apparently concerned with thermodynamics, it is much more general than its name implies and might well be called the second law of physics, or indeed, of nature.

We begin the study of the second law by considering some common processes from everyday life, as illustrated in Fig. 3.4.1. In each diagram, time is increasing towards the right, as shown at the bottom of the diagram. Each process has the common feature that if a cine film had been taken of the process, and of any measurements necessary to indicate temperature, electric charge, etc. and if the film was later projected backwards, the result would be sheer disbelief. All normal people viewing such a film would know,

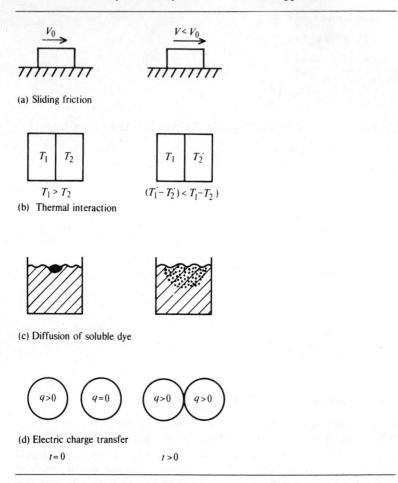

(a) Sliding friction

(b) Thermal interaction

(c) Diffusion of soluble dye

(d) Electric charge transfer

Fig. 3.4.1. Processes subject to the second law.

beyond doubt, that what they saw projected had been subjected to *time reversal* and that there was no possibility of the projection representing real processes. It is the second law of thermodynamics that determines the direction in which the projector must be operated to produce a representation of the real world and the processes taking place in it. The second law is a forbidding law that is held to be inviolate beyond any measure of dispute.

There are many ways of expressing the second law as Fig. 3.4.1 suggests, but for our purposes here it will be sufficient to quote a version of it that is particularly appropriate to thermal processes, in other words the process of example (b) in Fig. 3.4.1. Thus we may say that *in the process of energy transmission by heating, there is an energy transfer from the substance at the higher temperature to that at the lower temperature, and never in the reverse direction.* A popular version of this, and perfectly correct too, states that 'the hotters get cooler and the coolers get hotter!'

In common with other laws of thermodynamics it is necessary to be able to enlarge on the statement, and if possible it is convenient to express it in the form of a mathematical relationship. One possible way is to invoke equation [2.3.1] for the relationship between thermal flux and temperature gradient, i.e.

$$\dot{q} = -\kappa \frac{\mathrm{d}T}{\mathrm{d}x}$$

Here $\dot{q}$ is the rate of energy transfer in the OX direction and $\mathrm{d}T/\mathrm{d}x$ the temperature gradient in that same direction. The second law implies that for $\dot{q}$ positive it is necessary that $\mathrm{d}T/\mathrm{d}x$ is negative, and will be satisfied providing the thermal conductivity κ is never a negative quantity. Thus to avoid violating the second law κ *must* always be positive or zero. However such a statement of the second law has a severe drawback since not all materials have a constitutive equation that can be represented by equation [2.3.1]. An alternative approach is therefore necessary to avoid complexity with various constitutive equations. The key to this is to examine changes in *entropy*, the thermodynamic property introduced briefly in section 2.2 and defined by equation [2.2.21].

$$T\delta s = \delta u + p\delta v$$

where δs is the change in entropy s.

It was shown then that by using the equation of thermal interaction δs may be expressed in terms of ΔQ and $\Delta \Phi$, giving

$$T\delta s = \Delta Q + \Delta \Phi \qquad\qquad [3.4.1]$$

We now use this equation to assess the change in entropy of the systems undergoing a thermal interaction, as shown in Fig. 3.4.2.

In Fig. 3.4.2 we have two large systems S and R at temperatures T_S, T_R with $T_S \geqslant T_R$ and in thermal contact by means of a poor

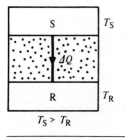

Fig. 3.4.2. Thermal interaction between systems S and R.

thermal conductor, so that each system may be considered to be at a sensibly uniform temperature which changes little over an element of positive thermal energy transfer $\Delta Q'$. There is no motion in this thermal process and therefore $\Delta\Phi = 0$, giving from equation [3.4.1] for each system

$$T\delta s = \Delta Q \qquad\qquad [3.4.2]$$

Note that in equation [3.4.1] ΔQ is an energy transfer per unit mass and hence s is an *entropy* per unit mass. However we now wish to assess values of *absolute* changes in entropy, but note that in equation [3.4.2] it is immaterial whether δs and ΔQ are absolute values or are on a per unit mass basis. We shall here take δS and $\Delta Q'$ to be the absolute values. We have from equation [3.4.2] for each system

$$T_S \delta S_S = -\Delta Q' \quad \text{or} \quad \delta S_S = -\frac{\Delta Q'}{T_S}$$

$$T_R \delta S_R = \Delta Q' \quad \text{or} \quad \delta S_R = +\frac{\Delta Q'}{T_R}$$

There is no corresponding change in entropy for the poor conductor if the system is in a steady state. We find, therefore, that

$$\delta S = (\delta S_S + \delta S_R) = \Delta Q'\left(\frac{1}{R_R} - \frac{1}{T_S}\right)$$

However, $T_S \geqslant T_R$ and $\Delta Q' \geqslant 0$ and thus

$$\delta S = (\delta S_S \delta S_R) \geqslant 0 \qquad\qquad [3.4.3]$$

Thus, a direct consequence of the second law is that, in the thermal interaction of the two systems, the total entropy of the systems has increased. It is a simple extension to now consider thermal interaction for more complex situations, but inevitably the resultant total change in entropy must be positive. If S is the total entropy of systems undergoing thermal interactions, then we have

$$\delta S \geqslant 0 \qquad\qquad [3.4.4]$$

Equation [3.4.4] is a mathematical statement of the *second law of thermodynamics*. It has been deduced in this analysis for the particular process of thermal interaction under a temperature gradient or difference, but in fact is *generally* true for all processes, providing the entropy is appropriately defined. It follows from this that if we now consider a process in which there is frictional dissipation $\Delta\Phi$, but where $\Delta Q = 0$ then we have, using equation [3.4.1],

$$\delta s = \frac{\Delta\Phi}{T} \geqslant 0 \qquad\qquad [3.4.5]$$

Since T can only take positive values, then the second law dictates that the frictional dissipation $\Delta\Phi$ is necessarily positive. This confirms the results of the analysis in section 2.4 for a Newtonian fluid and in fact proves conclusively that the coefficient of viscosity μ can never take negative values for real fluids.

Reversibility and the Carnot theorem

It is now clear that the non-reversibility of certain physical processes is a consequence of these processes being involved with a net increase in entropy of all the systems associated with the process. A process can only be considered reversible if the total entropy does not change during the process. A necessary requirement for this is the complete absence of frictional dissipation. For thermal processes to be reversible it is necessary, as shown by the analysis above, that cooling or heating processes are accomplished without significant temperature difference. It is only if these conditions are satisfied that a process or a series of processes can be considered to be in any sense reversible. It proves to be impossible in the real world to achieve complete reversibility, but despite this it is useful to consider some of the consequences of arranging for a work output device – or engine – to work in a reversible manner.

We begin by restricting our analysis to engines which operate to a cycle, which may have a number of component processes, but where at the end of each cycle the working medium is returned to its original condition of internal energy, pressure, etc. Integration of any variable over the cycle we denote by $\oint$ and we note, in passing, that $\oint d\psi$ where ψ is any material property, is necessarily zero whereas $\oint \Delta W$ or $\oint \Delta Q$ is not. Cyclic integration of a compound function, i.e. $\oint p\, dv$ again is not necessarily zero. We consider first the momentum equation [1.4.15] which, with the inverse density replaced by specific volume, is

$$\delta\left(\frac{v^2}{2}\right) + v\delta p + \Delta\Phi + \Delta W + \delta\phi = 0$$

giving

$$\oint d\left(\frac{v^2}{2}\right) + \oint v\, dp + \oint \Delta\Phi + \oint \Delta W + \oint d\phi = 0$$

but

$$\oint d\left(\frac{v^2}{2}\right) = 0, \qquad \oint d\phi = 0$$

and hence

$$\oint v\, dp + \oint \Delta\Phi + \oint \Delta W = 0$$

or

$$\oint \Delta W = -\oint v \, \mathrm{d}p - \oint \Delta \Phi \qquad [3.4.6]$$

From the second law we have $\oint \Delta \Phi \geqslant 0$, and thus it is apparent from equation [3.4.6] that, for a work output device, the effect of friction is to reduce the work output from the corresponding output of a frictionless or reversible engine. The first law or principle of conservation of energy requires that the net energy supplied to the system must be zero over the cycle and hence

$$\oint \Delta W = \oint \Delta Q \qquad [3.4.7]$$

regardless of the magnitude of $\oint \Delta \Phi$.

A useful way to present the details of the processes in an engine cycle, when changes in kinetic and potential energy can be ignored, is in the form of a plot of pressure against specific volume, usually called a pv diagram, as shown diagramatically in Fig. 3.4.3.

In the diagram the processes (3–4) and (4–1) are where work is being done on the medium and in processes (1–2) and (2–3) there is a work output from the system. The net work *output* for a frictionless cycle is from equation [3.4.6]

$$\oint \Delta W = -\oint v \, \mathrm{d}p = \oint p \, \mathrm{d}v$$

   = area within the closed curve on the pv diagram.

If the cycle involves any degree of frictional dissipation then the work output is, as equation [3.4.6] shows, less than this value. A diagram is therefore a useful method of indicating the maximum possible work output from an engine cycle.

If the process cycle shown in Fig. 3.4.3 is to be reversible it is not sufficient just to consider it to be frictionless. All energy transfers by heating must be at constant temperature to avoid the otherwise inescapable production of entropy. Only two processes are possible

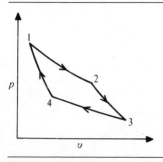

Fig. 3.4.3. The pressure/volume or pv diagram.

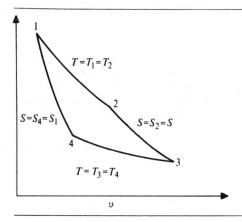

Fig. 3.4.4. The representation of a cycle on a *pv* diagram.

in a reversible cycle in which a net heat input is used to produce an equal work output. These are firstly an adiabatic, frictionless (i.e. isentropic) process, and secondly a constant temperature heating or cooling which must also be frictionless. Most heating or cooling processes take place with little if any motion and can thus be considered frictionless.

It turns out, as might be expected, that a minimum of four processes is required to produce a reversible working cycle and that this must be done by taking each of the two possible reversible processes in turn. The working cycle for a reversible engine is shown in Fig. 3.4.4. Such a cycle is called a *Carnot cycle*, after the French mining engineer Sadi Carnot. It is quite impossible to construct an engine which is reversible, but despite this a study of the cycle will prove invaluable.

In the cycle of Fig. 3.4.4, the path (1–2) is a constant-temperature heat supply and (3–4) a constant-temperature heat rejection. The two remaining paths (2–3) and (4–1) are both isentropic processes. We denote the heat supply and rejection by Q_S and Q_R and use W_0 for the net work output. From equation [3.4.7] we have

$$\oint \Delta W = \oint \Delta Q$$

or

$$W_0 = Q_S - Q_R \qquad [3.4.8]$$

In a practical engine the heat supply Q_S must be paid for and the useful output is W_0; the heat rejection Q_R, in general, being of little value. Consequently, economic interest is directed at the ratio between what can in principle be sold, W_0, and what has to be paid for, Q_S. The ratio (W_0/Q_S) is called the *conversion ratio*, or sometimes the *thermal efficiency* and denoted by η. The ratio (Q_R/Q_S) is correspondingly called the *rejection ratio* ρ. Using equation [3.4.8] it is easy to show that $\eta = (1 - \rho)$.

An important development can now be made by comparing the conversion ratio of a reversible engine with another engine of unspecified type when both are operating between the same fixed temperature limits. We consider now Fig. 3.4.5(a), in which engine A is a reversible engine operating to the Carnot cycle and engine B is another engine. We assume that engine B has a higher conversion ratio than A, or in other words $\eta_B > \eta_A$. Furthermore, we arrange that the size of the two engines is such that the power outputs are equal or $W_A = W_B$. Thus we have

$$\eta_A = \frac{W}{Q_1}, \qquad \eta_B = \frac{W}{Q_3}$$

but

$$\eta_B > \eta_A \quad \text{and} \quad W = Q_1 - Q_2 = Q_3 - Q_4$$

and thus

$$Q_1 > Q_3 \quad \text{and} \quad Q_2 > Q_4 \tag{3.4.9}$$

Now engine A is reversible and can be operated in the reversed mode, as shown in Fig. 3.4.5(b), with all the energy transfers retaining the same magnitudes but reversed in direction. Furthermore, in the reversed mode the power input can be provided exactly by engine B. However, if we consider the situation presented by this system, we see that the inequalities [3.4.9] show clearly that there is a net energy transfer by heating from the lower temperature T_R to the higher T_S, a situation forbidden by the second law. Consequently, it follows that the initial assumption that $\eta_B > \eta_A$ is untenable and in fact we have proved that $\eta_B \not> \eta_A$. Using the second law we have therefore proved that *no engine can be more efficient than a reversible engine when operating between the same upper and lower*

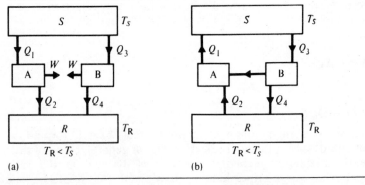

Fig. 3.4.5. Engines operating between the same temperature limits.

temperatures. This is known as the *Carnot theorem.* It was first proposed by Sadi Carnot and published in a paper on the subject in 1824. An easily proven corollary of the theorem is that any two reversible engines must have the same conversion ratio when operating between the same temperature limits.

The thermodynamic temperature scale

The concept of a reversible engine can now be used to define an absolute scale of temperature quite independent of any material properties. As shown in Fig. 3.4.6 we consider a number of reversible engines operating between thermal reservoirs S_1, S_2, S_3... at different temperatures, and a common rejection sink. For each engine the conversion ratio η is well defined, independently of any material properties, and we have

$$\eta_1 = \left(1 - \frac{Q_1'}{Q_1}\right), \qquad \eta_2 = \left(1 - \frac{Q_2'}{Q_2}\right), \qquad \eta_3 = \left(1 - \frac{Q_3'}{Q_3}\right) \text{etc.}$$

We know that if S_1 and S_2 are at the same degree of hotness then $\eta_1 = \eta_2$ and hence

$$\left(\frac{Q_1'}{Q_1} \cdot \frac{Q_2}{Q_2'}\right) = 1,$$

and consequently we can use $\{(Q_1'/Q_1) \cdot (Q_2/Q_2')\}$ as a measure of the ratio between the hotness of S_1 and S_2. There are many ways in which a temperature scale can now be made, of which perhaps the best is to arrange that the scale θ is such that $Q_1'/Q_1 = \theta_1'/\theta_1$ where θ_1 and θ_1' are the temperatures of S_1 and the rejection sink. Thus we have

$$\eta_1 = \left(1 - \frac{Q_1'}{Q_1}\right) = \left(1 - \frac{\theta_1'}{\theta_1}\right) \qquad [3.4.10]$$

Such a scale of temperature, independent of material properties, is

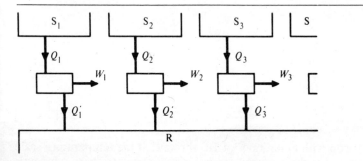

Fig. 3.4.6. Reversible engines operating between different temperature limits.

called an *absolute thermodynamic scale* and it remains now to show that one such scale is the perfect gas scale.

We do this by assuming that the reversible engine cycle shown previously in Fig. 3.4.4 is using a perfect gas as the working medium, and we analyse the individual processes in the cycle on this basis. All processes are without frictional dissipation and hence from the equation of thermal interaction we have

$$\Delta Q = \delta u + p\,\delta v$$

Process (1–2) is a constant-temperature heat supply and hence $\delta u = 0$, giving

$$Q_{12} = \int_1^2 \Delta Q = \int_1^2 p\,\mathrm{d}v = \int_1^2 RT\frac{\mathrm{d}v}{v} = RT\ln\left(\frac{v_2}{v_1}\right)$$

Similarly,

$$Q_{34} = RT'\ln\left(\frac{v_4}{v_3}\right)$$

or heat rejected is

$$Q' = -Q_{34} = RT'\ln\left(\frac{v_3}{v_4}\right)$$

Therefore

$$\eta = \left(1 - \frac{Q_1'}{Q_1}\right) = 1 - \frac{RT'\ln(v_3/v_4)}{RT\ln(v_2/v_1)}$$

However, since the processes (2–3) and (4–1) are adiabatic and frictionless we have from equation [2.2.20] that

$$\frac{T_3}{T_2} = \frac{T'}{T} = \left(\frac{v_2}{v_3}\right)^{\gamma-1} \quad\text{and}\quad \frac{T_4}{T_1} = \frac{T'}{T} = \left(\frac{v_1}{v_4}\right)^{\gamma-1}$$

Hence

$$\frac{v_2}{v_3} = \frac{v_1}{v_4} \quad\text{or}\quad \frac{v_2}{v_1} = \frac{v_3}{v_4}$$

consequently,

$$\eta = \left(1 - \frac{T'}{T}\right) \qquad\qquad [3.4.11]$$

Comparison with equation [3.4.10] shows that the temperature scale used for T, i.e. the perfect gas scale, *is* an absolute thermodynamic scale of temperature.

Example 3.4.1
A gas-turbine engine has a maximum flame-temperature
limit of 1300 K and has an air inlet temperature of 300 K. If
the thermal input in the combustion chamber is 4 MW
estimate a never-to-be-achieved maximum power output for
the engine, and the corresponding minimum heat rejection.

Data
$T_{max} = 1300$ K
$T_{min} = 300$ K
$Q_S = 4$ MW

Analysis
For an engine operating to a Carnot cycle we have, for the
conversion ratio η_c

$$\eta_c = \left(1 - \frac{Q_R}{Q_S}\right) = \left(1 - \frac{\dot{Q}_R}{\dot{Q}_S}\right) = \left(1 - \frac{T_{min}}{T_{max}}\right)$$

which gives η_c for temperature limits T_{max} and T_{min}.
   However, the actual conversion ratio of the engine $\eta \leqslant \eta_c$,
and thus η_c is a never-to-be-achieved maximum, since no
practical engine can be reversible. The value for $\dot{W} =$
$(\dot{Q}_S - \dot{Q}_R)$ can then be deduced.

Calculation

$$\eta_c = \left(1 - \frac{300}{1300}\right) = \frac{\dot{W}_c}{\dot{Q}_S}$$

Therefore

$$\dot{W}_c = \frac{4 \times 1000}{1300}$$
$$= 3\cdot08 \text{ MW}$$

Hence, for the actual engine,

$\dot{W} < 3\cdot08$ MW

The corresponding *minimum* heat rejection to the atmosphere is $\dot{Q}_R$, where

$\dot{Q}_R = (\dot{Q}_S - \dot{W}) = (4 - 3\cdot08)$
$\quad = 0\cdot92$ MW

The use of the Carnot theorem, together with equation
[3.4.11] for the conversion ratio, can prove extremely useful
in certain problems where estimates of maximum or
minimum values are required as in Example 3.4.1.

Example 3.4.2

A ship power plant is designed for a boiler outlet temperature of 700 K and uses sea-water to provide cooling in the
condenser. Estimate the increase in fuel consumption when
the ship sails from Scotland to the Azores. Assume average
sea-water temperature in Scottish waters to be 10 °C and
around the Azores 20 °C.

Data

$T_{max} = 700$ K

$T_{min} = 10$ °C $= 283$ K (Scotland)

$\qquad = 20$ °C $= 293$ K (Azores)

Analysis

Make an initial analysis using the Carnot cycle as in Example 3.4.1, and assume that the power requirement $\dot{W}$ is the
same throughout the voyage. We have then

$$\eta = \frac{\dot{W}}{\dot{Q}_S} = \left(1 - \frac{T_{min}}{T_{max}}\right)$$

Therefore

$$\dot{Q}_S = \dot{W}\left(\frac{T_{max}}{T_{max} - T_{min}}\right)$$

Fuel consumption is proportional to $\dot{Q}_S$ and hence the increase can be found. Percentage increase for the actual engine is likely to be of the same order of magnitude.

Calculation

(Scotland) $\dot{Q}_S = \dfrac{\dot{W} \times 700}{(700 - 283)}$

$\qquad\qquad = 1{\cdot}68\,\dot{W}$

(Azores) $\dot{Q}_S = \dfrac{\dot{W} \times 700}{(700 - 293)}$

$\qquad\qquad = 1{\cdot}72\,\dot{W}$

Hence percentage increase in fuel consumption is

$$\frac{(1{\cdot}72 - 1{\cdot}68)}{1{\cdot}68} \times 100 = \mathbf{2{\cdot}38\%}$$

3.5 Uncertainty and entropy

In section 3.4 the second law of thermodynamics was introduced
and its application to work and heat processes investigated in detail.

However, the examples illustrated in Fig. 3.4.1 show that there must be some more general method of expressing the idea behind the second law so that it may be applied to processes other than those analysed in section 3.4. If this can be done it should be possible to use it to rule out impossible processes that cannot conceivably take place in our material world.

The key to a more general statement of the second law and more widespread application is the property *entropy* which was first introduced in section 2.2 and defined by equation [2.2.21], namely

$$T\delta s = \delta u + p\,\delta v$$

where δs is the change in entropy per unit mass.

We now introduce a more general concept of entropy, and first of all lay down a set of rules which must be satisfied if it is to serve a useful purpose as a property of a body or a system. Mass, momentum, charge and energy are familiar properties of systems and bodies having the common feature that they are *extensive*, i.e. the mass of a large system is the sum of the masses of the individual components. We take the new property, entropy, also to be extensive. Furthermore, we associate entropy with the direction in which processes may be allowed to proceed by stating that for *any* interactions taking place in an isolated system, the entropy of the system can never decrease with time, i.e.

$$\left(\frac{dS}{dt}\right)_{\text{isolated system}} \geqslant 0 \qquad\qquad [3.5.1]$$

Here S is the absolute entropy of the system.

This equation is taken to be the general statement of the second law, applicable to any process that we can imagine taking place. This general statement applies equally well to biological phenomena, games of chance and electromagnetic or chemical processes, and it is assumed in each case that it will prove possible to arrive at a suitable definition for the entropy of the system being considered. The previous definition given in equation [2.2.21] is but one particular example appropriate to a system having only work and heat interactions.

We consider now a particular system shown in Fig. 3.5.1 consisting of a small round ball or marble in a round box that has nine radial dimples impressed into the base around a central column, and into which the ball can roll and then remain relatively immune to small disturbances of the box. The ball is initially placed in one of the dimples and then at intervals of time the box is given a small vertical shake which is just sufficient to dislodge the ball from its position in the dimple. We assign numbers 1, 2 ... 9 to the rows of dimples with 1 for the initial position of the ball, and at any given time we assign probabilities $p_1, p_2 \ldots p_9$ to the likelihood of the ball being in a particular dimple. The probabilities are scaled so that in

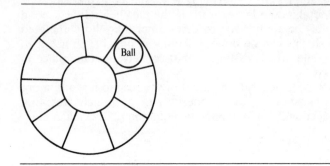

Fig. 3.5.1. Ball in dimpled box.

the usual convention $\sum_1^9 p_i = 1$. Our assignment of probabilities after each shake might look something like Table 3.5.1 if we assume that at each shake the ball has an equal chance of remaining where it is or moving to an adjacent dimple.

Eventually, as Table 3.5.1 indicates, the trend would be for the probability distribution to become more uniform, and after a very large number of shakes the distribution would be just that with all positions being equally probable. What is happening in this experiment is that as time proceeds our *uncertainty* about the position of the ball is increasing and eventually reaches a maximum when, in this particular example, all rows or *states* are equally probable. What is required is some measure of the uncertainty or peakiness of the probability distribution. To be useful it must be a *single* parameter rather than a large number of individual probabilities, and to be of general use it must be *extensive* so that it is possible to describe the uncertainty of a large system in terms of the sum of the individual uncertainties of the smaller systems within it. The measure of uncertainty is called the *entropy S*, and is defined in terms of

Table 3.5.1

Position	1	2	3	4	5	6	7	8	9
Start	1	0	0	0	0	0	0	0	0
Shake 1	1/3	1/3	0	0	0	0	0	0	1/3
2	3/9	2/9	1/9	0	0	0	0	1/9	2/9
3	7/27	6/27	3/27	1/27	0	0	1/27	3/27	6/27
4	19/81	16/81	10/81	4/81	1/81	1/81	4/81	10/81	16/81
5	51/243	45/243	30/243	15/243	6/243	6/243	15/243	30/243	45/243
6									
7									
8									
9									
n (large)	1/9	1/9	1/9	1/9	1/9	1/9	1/9	1/9	1/9

the probabilities p_i for the n possible outcomes by the equation

$$S = -k \sum_{i=1}^{n} p_i \ln p_i \qquad [3.5.2]$$

Here k is the Boltzmann constant and has the value $k = 1\cdot380 \times 10^{-23}$J/K. This general definition of entropy is known as the Gibbs equation and applies to all systems regardless of whether the process taking place is in an initial or relatively settled state of development. It is interesting now to compute the values of S for the stages of development for our system of the ball in the box. The results are as follows:

State	$S(k)$
Initially	0
Shake 1	1·099
2	1·522
3	1·750
4	1·902
5	1·996

After each shake the entropy has increased and, after a very large number, its value finally tends towards

$$S = -k \sum_{i=1}^{n} p_i \ln p_i = k \ln\left(\frac{1}{9}\right) = 2\cdot20\,k$$

The values of S are shown in Fig. 3.5.2.

In the experiment discussed here it was obvious that ultimately the ball was equally likely to be in any row in the box, but this need

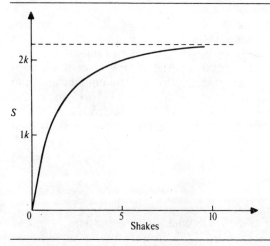

Fig. 3.5.2. Entropy values after shaking.

not be so in the general case. For a general process there may be other constraints which have to be taken into account, and in these the final or maximum entropy can only be found by maximising S subject to whatever constraints apply to the system or process. When S has achieved this maximum value the system is said to be in *equilibrium* and the probability distribution is the *equilibrium distribution*.

For a simple monatomic gas it is possible to use equation [3.5.2] for S and compute values for the entropy, in terms of its energy and density. However, a general substance is far too complex in its molecular structure and interactions for this to be done and we have to adopt a macroscopic rather than a microscopic approach. To do this we shall use the principle of the second law that in any process the entropy of the system cannot decrease and that the equilibrium state – where no interaction or change in the probability distribution is taking place – is the state of *maximum* entropy or uncertainty.

Earlier, in section 2.2, we introduced what was called the two-property rule for the thermodynamic properties p, v, T, u and h and to which list we can now add the specific entropy s or entropy per unit mass. Formally, the rule states that at *equilibrium state*, and in a single phase, only two properties can be considered to be independent and any other is functionally related to the other two. Thus we have, for example

$$s = s(u, v)$$

or, in terms of absolute rather than specific quantities,

$$S = S(U, V) \qquad [3.5.3]$$

and the problem now is to determine the functional relationship of equation [3.5.3] in terms of known parameters.

We begin by considering the thermal interaction of two components A and B of an isolated system. We assume that the interaction results in no volume change of either A or B and thus there is no *mechanical* interaction. The total internal energy U of the system does not change as the system interacts towards an equilibrium state and when this is achieved we assign αU and $(1 - \alpha)U$ to be the equilibrium internal energies of A and B respectively. We have, then, that the entropy for the system S is the sum of the individual entropies $(S_A + S_B)$, i.e.

$$\begin{aligned}
S &= S_A + S_B \\
&= S(U, V)_A + S(U, V)_B \\
&= S(\alpha U, V)_A + S((1 - \alpha)U, V)_B
\end{aligned}$$

The parameters U, V_A and V_B are constant, independent of α, and hence the only variable is α itself. Consequently, the equilibrium state is that for which $S = S(\alpha)$ has a maximum value. Therefore, at equilibrium

$$\frac{dS}{d\alpha} = \left(\frac{\partial S_A}{\partial U_A}\right)_{V_A} \cdot \frac{\partial U_A}{\partial \alpha} + \left(\frac{\partial S_B}{\partial U_B}\right)_{V_B} \cdot \frac{\partial U_B}{\partial \alpha} = 0$$

but

$$\frac{\partial U_A}{\partial \alpha} = U \quad \text{and} \quad \frac{\partial U_B}{\partial \alpha} = -U$$

and hence for the maximum conditions

$$\left(\frac{\partial S_A}{\partial U_A}\right)_{V_A} = \left(\frac{\partial S_B}{\partial U_B}\right)_{V_B} \qquad [3.5.4]$$

Thus, thermal equilibrium between A and B is achieved when the quantity $(\partial S/\partial U)_V$ is the same for both, and consequently $(\partial S/\partial U)_V$ is a monatonic function of the absolute thermodynamic temperature T. The functional relationship between $(\partial S/\partial U)_V$ and T is not arbitrary if the entropy S is to be consistent with the definition in terms of probabilities given in equation [3.5.2]. It can be shown that for consistency

$$\left(\frac{\partial S}{\partial U}\right)_V = \frac{1}{T} \qquad [3.5.5]$$

where T is the temperature measured on the perfect gas scale.

We have now reached an interesting position for we have from equation [3.5.3],

$$S = S(U, V)$$

or

$$\delta S = \left(\frac{\partial S}{\partial U}\right)_V \delta U + \left(\frac{\partial S}{\partial V}\right)_U \delta V$$

$$= \frac{\delta U}{T} + \left(\frac{\partial S}{\partial V}\right)_U \delta V \qquad [3.5.6]$$

which shows that if we can now perform a similar analysis on the function $(\partial S/\partial V)_U$ we shall have arrived at a suitable equation for evaluating changes in entropy S.

We now consider again the interaction of the systems A and B, but this time relax the condition that V_A and V_B remain constant, to the less restrictive condition that $(V_A + V_B)$ remains constant. The two systems in coming to equilibrium can now have both a heat interaction and a work interaction caused by pressure difference between and volume change of the two systems A and B. Again, we let the final internal energies U_A and U_B be αU and $(1-\alpha)U$ respectively, and the corresponding volumes V_A and V_B be βV and $(1-\beta)V$

respectively. We now have two independently variable parameters α and β which necessitates two conditions for S to be a maximum. These are

$$\left(\frac{\partial S}{\partial \alpha}\right)_\beta = 0 \quad \text{and} \quad \left(\frac{\partial S}{\partial \beta}\right)_\alpha = 0$$

giving, by a similar analysis to that above,

$$\left(\frac{\partial S_A}{\partial U_A}\right)_{V_A} U = \left(\frac{\partial S_B}{\partial U_B}\right)_{V_B} U$$

and

$$\left(\frac{\partial S_A}{\partial V_A}\right)_{U_A} V = \left(\frac{\partial S_B}{\partial V_B}\right)_{U_B} V$$

or

$$\left.\begin{array}{l} \left(\dfrac{\partial S_A}{\partial U_A}\right)_{V_A} = \left(\dfrac{\partial S_B}{\partial U_B}\right)_{V_B} \\[3mm] \text{and} \\[3mm] \left(\dfrac{\partial S_A}{\partial V_A}\right)_{U_A} = \left(\dfrac{\partial S_B}{\partial V_B}\right)_{U_B} \end{array}\right\} \qquad [3.5.7]$$

for S to be a maximum, i.e. the two systems to be at equilibrium. The first of these conditions is simply that of thermal equilibrium discussed above, and the second is the condition required for mechanical equilibrium.

Mechanical equilibrium cannot be achieved when V_A and V_B can vary if the pressures p_A and p_B are different and thus $(\partial S/\partial V)_U$ is a function of pressure and possibly also temperature, since the conditions of equations [3.5.7] incorporate thermal equilibrium. The dimensions of S are, from equation [3.5.5], energy/unit temperature and thus the dimensions of $(\partial S/\partial V)_U$ are of energy/unit temperature volume or pressure/unit temperature. Thus we have

$$\left(\frac{\partial S}{\partial V}\right)_U = \frac{p}{T} \qquad [3.5.8]$$

is a possible relationship and can indeed be shown to be the only satisfactory relationship that is consistent with the normal definition of pressure. Substituting for $(\partial S/\partial V)_U$ in equation [3.5.6] gives finally

$$\delta S = \frac{\delta U}{T} + \frac{p}{T}\delta V$$

or in terms of specific quantities

$$\delta s = \frac{\delta u}{T} + \frac{p}{T}\,\delta v \qquad\qquad [3.5.9]$$

Equation [3.5.9] is known as the *Gibbs equation* for a simple compressible substance, and is but one of a series of Gibbs equations that can be developed for more complex interactions than the heat and work interactions considered here.

The Gibbs equation enables changes in entropy to be calculated from experimental data for any substance and is used in the preparation of thermodynamic tables. For a perfect gas equation [3.5.9] can be developed further to produce even simpler expressions for entropy change. For a perfect gas then, we have

$$\delta s = \frac{\delta u}{T} + \frac{p\,\delta v}{T}$$

$$= \frac{c_v \delta T}{T} + \frac{R}{v}\,\delta v$$

and hence

$$s_2 - s_1 = c_v \int_1^2 \frac{\mathrm{d}T}{T} + R \int_1^2 \frac{\mathrm{d}v}{v}$$

$$= c_v \ln\left(\frac{T_2}{T_1}\right) + R \ln\left(\frac{v_2}{v_1}\right) \qquad\qquad [3.5.10]$$

or

$$s_2 - s_1 = c_v \ln\left(\frac{p_2 v_2}{p_1 v_1}\right) + R \ln\left(\frac{v_2}{v_1}\right)$$

$$= c_v \ln\left(\frac{p_2}{p_1}\right) + c_p \ln\left(\frac{v_2}{v_1}\right) \qquad\qquad [3.5.11]$$

This last equation for $(s_2 - s_1)$ is particularly useful and instructive when rearranged in the form

$$s_2 - s_1 = c_v \ln\left(\frac{p_2 v_2^\gamma}{p_1 v_1^\gamma}\right) \qquad\qquad [3.5.12]$$

where $\gamma = (c_p/c_v)$ is the specific heat ratio. In this form it can be used as a quick check on experimental data, particularly for adiabatic flow. If the flow direction is from $(1 \to 2)$, then we must have that $s_2 > s_1$ and this can only be satisfied if $p_2 v_2^\gamma > p_1 v_1^\gamma$.

3.6 Dissipation and availability

The second law and the Carnot cycle have already introduced us to the situation that devices constructed to produce a work output from a thermal input can only do this with a conversion ratio less than unity, and furthermore the effect of frictional dissipation and irreversibility is to reduce this conversion ratio even further. The analysis of the operating cycles of practical power plants in the next chapter will show that the overall conversion ratio depends critically upon the irreversibilities within the components of the plant, for example boilers, turbines, etc. The intention in this section is to look a little further into these irreversibilities and to develop a methodology for assessing the effect of them upon the overall operation of thermodynamic systems and processes.

We consider the steady-flow system shown in Fig. 3.6.1. In this system there is a heat transfer from the surrounding environment of $\dot{Q}$ and the system gives a work output of $\dot{W}$. The temperature of the environment is taken to be T_0. If we ignore changes in kinetic and potential energy the energy balance for the system is

$$\dot{W} - \dot{Q} = \dot{M}(h_1 - h_2) \qquad [3.6.1]$$

We now consider the rate at which the system produces entropy $\dot{S}$. This may be done most conveniently by equating the production of entropy to the transfer of entropy to the system through the mass fluxes and the heat transfer.

The rate at which entropy is transferred out of the system by the mass flow is

$$\dot{M}(s_2 - s_1) \qquad [3.6.2]$$

For the heat transfer $\dot{Q}$ we have from equation [2.2.22] that the entropy transferred per unit mass in a heating process at temperature T_0 is given by

$$\delta S = \frac{\delta Q}{T_0} \qquad [3.6.3]$$

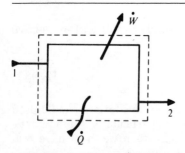

Fig. 3.6.1. Steady flow system.

and thus for a steady continuous process the *rate* at which entropy is transferred is

$$\dot{S} = \frac{\dot{Q}}{T_0}$$

We can now equate the entropy transfers across the boundaries of the system and obtain $\dot{S}$ which is then given by

$$\dot{S} = \dot{M}(s_2 - s_1) - \frac{\dot{Q}}{T_0} \qquad [3.6.4]$$

We now use equation [3.6.1] to eliminate $\dot{Q}$ from this equation for the entropy production rate, and obtain

$$\dot{S} = \dot{M}(s_2 - s_1) + \frac{\dot{M}}{T_0}(h_1 - h_2) - \frac{\dot{W}}{T_0} \qquad [3.6.5]$$

The second law of thermodynamics requires that $\dot{S} \geqslant 0$, and thus for this to be satisfied we must have

$$\dot{W} \leqslant \dot{M}\{(h_1 - T_0 s_1) - (h_2 - T_0 s_2)\} \qquad [3.6.6]$$

It follows from this result that the value obtained for the work output by using the equality embodied in equation [3.6.6] is a *maximum* work output. This maximum work output represents the available work output that may be achieved from the given system surrounded by an environment at temperature T_0. The term $(h - T_0 s)$ is known as the *steady flow availability* function, and, unlike the previous thermodynamic functions, depends upon both fluid properties *and* the environmental temperature.

Example 3.6.1
An air heater consists of a series of tubes over which atmospheric air is passed and through which steam is condensed. The inlet and outlet conditions for the steam are $p_1 = p_2 = 1000$ kN/m^2 and $x_1 = 0.9$, $x_2 = 0.2$ and the mass

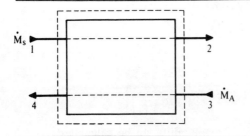

flow rate is 1·5 kg/sec. The air inlet pressure and temperature are 100 kN/m² and 285 K, and the outlet pressure and
mass flow rate are 100 kN/m² and 14·5 kg/sec respectively.
Find the loss or gain of availability for the air and steam
flows. Neglect kinetic and potential energy changes.

Data

$p_1 = p_2 = 1000$ kN/m²

$x_1 = 0·9$, $x_2 = 0·2$

$\dot{M}_S = 1·5$ kg/sec

$p_3 = p_4 = 100$ kN/m²

$T_3 = 285$ K

$\dot{M}_A = 14·5$ kg/sec

Analysis
Energy balance gives

$$\dot{M}_S(h_1 - h_2) = \dot{M}_A(h_4 - h_3) = \dot{M}_A c_p (T_4 - T_3)$$

Enthalpies h_1, h_2 are found from tables, and hence T_4 can
be found.

For the steam flow there is no work output and hence the
loss of availability is

$$\dot{W}_{Av} = \dot{M}_S\{(h_1 - T_0 s_1) - (h_2 - T_0 s_2)\}$$

A value for $\dot{W}_{Av}$ can now be found using tabulated values
of h_1, s_1, h_2, s_2, together with $T_0 = T_3 = 285$ K.

For the air flow we have $(h_3 - h_4) = c_p(T_3 - T_4)$ and
$(s_3 - s_4)$ can be found from equation [3.5.12], i.e.

$$s_3 - s_4 = c_v \ln\left(\frac{p_3 v_3^\gamma}{p_4 v_4^\gamma}\right)$$

$$= c_v \ln\left\{\frac{p_3}{p_4}\cdot\left(\frac{T_3}{p_3}\cdot\frac{p_4}{T_4}\right)^\gamma\right\}$$

$$= c_v \ln\left\{\left(\frac{p_4}{p_3}\right)^{\gamma-1}\cdot\left(\frac{T_3}{T_4}\right)^\gamma\right\}$$

$$= \left(\frac{c_p}{\gamma}\right)\left\{(\gamma-1)\ln\left(\frac{p_4}{p_3}\right) + \gamma \ln\left(\frac{T_3}{T_4}\right)\right\}$$

and hence the gain of availability can be calculated.

Thermodynamic table values for steam

$p = 1000 \text{ kN/m}^2$

$T_S = 453 \text{ K}$

$h_1 = 763 \text{ kJ/kg}, \qquad h_g = 2778 \text{ kJ/kg}$

$s_1 = 2{\cdot}138 \text{ kJ/kg K}, \qquad s_g = 6{\cdot}586 \text{ kJ/kg K}$

Calculation

$h_1 = (1 - 0{\cdot}9) \times 763 + 0{\cdot}9 \times 2778$

$\quad = \underline{2576 \text{ kJ/kg}}$

$h_2 = (1 - 0{\cdot}2) \times 763 + 0{\cdot}2 \times 2778$

$\quad = \underline{1166 \text{ kJ/kg}}$

Therefore

$$1{\cdot}5(2576 - 1166) = 14{\cdot}5 \times 1{\cdot}0(T_4 - 285)$$

and

$$T_4 = 431 \text{ K}$$

For steam flow

$s_1 = (1 - 0{\cdot}9) \times 2{\cdot}138 + 0{\cdot}9 \times 6{\cdot}586$

$\quad = \underline{6{\cdot}141 \text{ kJ/kg K}}$

$s_2 = (1 - 0{\cdot}2) \times 2{\cdot}138 + 0{\cdot}2 \times 6{\cdot}586$

$\quad = \underline{3{\cdot}028 \text{ kJ/kg K}}$

Hence

$$(\dot{W}_{Av})_S = 1{\cdot}5\{2576 - (285 \times 6{\cdot}141) - 1166 + (285 \times 3{\cdot}028)\}$$
$$= \underline{783{\cdot}9 \text{ kW}}$$

For air flow

$$s_3 - s_4 = \frac{1{\cdot}0}{1{\cdot}4}\left\{0{\cdot}0 + 1{\cdot}4 \ln\left(\frac{285}{431}\right)\right\} = \underline{-0{\cdot}414 \text{ kJ/kg K}}$$

Hence

$$(\dot{W}_{Av})_A = 14{\cdot}5\{1{\cdot}0(285 - 431) + 285 \times 0{\cdot}414\} = \underline{-407{\cdot}8 \text{ kW}}$$

There is thus a *gain* of availability for the air flow. The net loss of availability during the process is thus

$$(\dot{W}_{Av})_S + (\dot{W}_{Av})_A = 783{\cdot}9 - 407{\cdot}8$$
$$= \mathbf{376{\cdot}1 \text{ kW}}$$

There are a number of alternative ways in which an availability function can be defined, each one applicable to a particular process. The appropriate function enables the loss of availability in complex processes – occurring in, say, a process plant – to be evaluated for each stage in the process and enables the points of maximum loss to be highlighted and the process perhaps better organised.

An alternative approach for fluid machinery is to evaluate losses – or dissipative effects – by means of an efficiency. Taking as an example a compression process between pressures p_1 and p_2 we have for the work *input*, $\dot{W}_{12}$ from equation [1.4.16]

$$\dot{W}_{12} = \dot{M} \int_1^2 \frac{dp}{\rho} + \dot{\Phi}_{12} \qquad [3.6.7]$$

or

$$\dot{W}_{12} = \dot{M} \int_1^2 v \, dp + \dot{\Phi}_{12} \qquad [3.6.8]$$

Let us assume that the compression process (1–2) is as shown on the pv diagram of Fig. 3.6.2. From equation [3.6.8] we see that for the process path on the pv diagram the minimum work input to complete the path is $\dot{M} \int_1^2 v \, dp$, and thus we may define an efficiency for this compression process η_c by the relationship

$$\eta_c = \frac{(\dot{W}_{12})_{\min}}{\dot{W}_{12}} = \frac{\dot{M} \int_1^2 v \, dp}{\dot{M} \int_1^2 v \, dp + \dot{\Phi}_{12}} \qquad [3.6.9]$$

This definition of an efficiency is known as a *path efficiency* for the compression process and, although completely unambiguous, it suffers from the severe drawback that for evaluation of $\int_1^2 v \, dp$ we

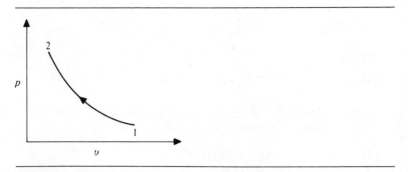

Fig. 3.6.2. A compression process on the pv diagram.

require knowledge of the relationship between p and v at each stage in the compression process. In practice, this information is generally not available and consequently we must adopt some more practical, if less ideal, method.

Many steady-flow compression and expansion processes are approximately adiabatic and thus from equation [3.6.1] the work done, $\dot{W}_{12}$, is given by

$$\dot{W}_{12} = \dot{M}(h_2 - h_1)$$

If the compression process was frictionless, then the process would be isentropic and the corresponding final enthalpy $h_{2_{is}}$ would be the enthalpy of the fluid at pressure p_2 corresponding to $s_2 = s_1$. For the isentropic process the work done $(\dot{W}_{12})_{is}$ is simply

$$(\dot{W}_{12})_{is} = \dot{M}(h_{2_{is}} - h_1)$$

and represents the work done in an adiabatic frictionless process between pressures p_1 and p_2. We can now define an isentropic efficiency for the adiabatic compression process by the relationship

$$(\eta_{is})_c = \frac{(\dot{W}_{12})_{is}}{\dot{W}_{12}} = \frac{\dot{M}(h_{2_{is}} - h_1)}{\dot{M}(h_2 - h_1)}$$

$$= \frac{(h_{2_{is}} - h_1)}{(h_2 - h_1)} \qquad [3.6.10]$$

For an expansion process the relationship is normally inverted so that $(\eta_{is})_e \leqslant 1$ and we then obtain

$$(\eta_{is})_e = \frac{(h_1 - h_2)}{(h_1 - h_{2_{is}})} \qquad [3.6.11]$$

where $h_{2_{is}}$ is now the enthalpy at pressure p_2 that would be obtained if the process had been adiabatic *and* frictionless.

Each of these isentropic efficiencies is easy to apply in practice since equations [3.6.10] and [3.6.11] depend only on the end points of the process and *not* on information at intermediate points, as in equation [3.6.9].

Example 3.6.2
Steam at a temperature and pressure of 400 °C and 4000 kN/m^2 is expanded adiabatically in a turbine to a pressure of 800 kN/m^2. The mass flow rate is 120 kg/sec and the isentropic efficiency of the expansion process is 0·87. Find the final condition of the steam and the work output of the turbine.

Data

$p_1 = 4000 \text{ kN/m}^2, \qquad p_2 = 800 \text{ kN/m}^2$

$T_1 = 400\,°C = 673 \text{ K}$

$\dot{M} = 120 \text{ kg/sec}$

$(\eta_{is})_e = 0{\cdot}87$

Analysis and calculation

From tables we have

$h_1 = 3214 \text{ kJ/kg}, \qquad s_1 = 6{\cdot}769 \text{ kJ/kg K}$

For an isentropic process (1–2′) between p_1 and p_2 we have

$s_2' = s_1 = 6{\cdot}769 \text{ kJ/kg K} \quad \text{at} \quad p_2 = 800 \text{ kN/m}^2$

thus from tables

$h_2' = \underline{2817 \text{ kJ/kg}}$

Hence

$$(\eta_{is})_e = 0{\cdot}87 = \frac{(h_1 - h_2)}{(h_1 - h_2')}$$

Therefore

$h_2 = h_1 - 0{\cdot}87(h_1 - h_2') = 3214 - 0{\cdot}87(3214 - 2817)$

$\quad = \underline{2868} \text{ kJ/kg}$

And thus from tables the final condition of the steam is

$p_1 = \mathbf{800 \text{ kN/m}^2}, \qquad T_2 = \mathbf{213\,°C = 486 \text{ K}}$

The work output is $\dot{W}_{12}$ where

$\dot{W}_{12} = \dot{M}(h_1 - h_2)$

$\quad = 120(3214 - 2868)$

$\quad = \mathbf{41\,520 \text{ kW}}$

When the working fluid is a perfect gas the isentropic efficiencies may be developed further by using the perfect gas relationship for an isentropic process, equation [2.2.19]. This gives for the isentropic process (1–2′),

$$\frac{T_2'}{T_1} = \left(\frac{p_2}{p_1}\right)^{(\gamma-1)/\gamma}$$

and thus

$$h_2' = c_p T_2' = c_p T_1 \left(\frac{p_2}{p_1}\right)^{(\gamma-1)/\gamma}$$

Substitution in equations [3.6.10] and [3.6.11] gives

$$(\eta_{is})_c = \frac{T_1\{(p_2/p_1)^{(\gamma-1)/\gamma}-1\}}{(T_2-T_1)} \qquad [3.6.12]$$

and

$$(\eta_{is})_e = \frac{(T_1-T_2)}{T_1\{1-(p_2/p_1)^{(\gamma-1)/\gamma}\}} \qquad [3.6.13]$$

Example 3.6.3
Air is compressed by an adiabatic steady-flow process from pressure $p_1 = 100 \text{ kN/m}^2$ to $p_2 = 900 \text{ kN/m}^2$. The initial temperature is 300 K and the mass flow rate 60 kg/sec. Measurements of the input power to the compressor give a working rate of 21 MW. Find the isentropic efficiency for the compression process, assuming little change in kinetic energy between inlet and outlet.

Data
$p_1 = 100 \text{ kN/m}^2, \qquad p_2 = 900 \text{ kN/m}^2$
$T_1 = 300 \text{ K}$
$\dot{M} = 60 \text{ kg/sec}$
$\dot{W} = 21 \text{ MW}$

Analysis
    Energy balance gives

$\dot{W} = \dot{M}c_p(T_2-T_1)$

$T_2 = T_1 + \dfrac{\dot{W}}{\dot{M}c_p}$

But from equation [3.6.12]

$(\eta_{is})_c = \dfrac{T_1\{(p_2/p_1)^{(\gamma-1)/\gamma}-1\}}{(T_2-T_1)}$

and thus $(\eta_{is})_c$ can now be found.

Calculation

$T_2 = 300 + \dfrac{21\,000}{60 \times 1\cdot 0}$

$\quad = 650 \text{ K}$

Therefore

$$(\eta_{is})_c = \frac{300\{(900/100)^{0.4/1.4} - 1\}}{(650 - 300)}$$

$$= \mathbf{0.749}$$

3.7 Third law

The availability function $(h - T_0 s)$ developed in equation [3.6.6] can only prove useful if it is possible to calculate or obtain absolute values of entropy s for the wide variety of substances that could be used in the processes we have discussed. We can already predict *changes* in entropy by using the Gibbs equation [3.5.9], but this tells us nothing about the absolute value of s. Examination of the three laws of thermodynamics discussed so far would show that there is no way to deduce from them any information on the absolute value of s for a particular substance, at known conditions of pressure and temperature.

The third law of thermodynamics, deduced in the early part of this century by Nernst, is concerned with the establishment of values for absolute entropies. Unlike the zeroth, first and second laws of thermodynamics, the third law is not firmly established as an inviolate law of nature. In most cases it would appear to work satisfactorily and is of unquestioned utility, but it is considered that there may be a number of circumstances under which it is not true. These will not be our concern here.

For a pure substance in equilibrium, the third law postulates that as the absolute temperature approaches zero, then so does the absolute entropy. In other words

$$\lim_{T \to 0} s = 0 \tag{3.7.1}$$

Combining this result with the Gibbs equation and integrating, we now obtain

$$s = \int_{T=0}^{T} \frac{du}{T} + \int_{T=0}^{T} \frac{p\,dv}{T} \tag{3.7.2}$$

and thus if we have knowledge, theoretical or experimental, of the variation of u, p and v along the path of any process approaching an absolute temperature of zero, we can compute an absolute value for the entropy of the substance. In our work here we shall not have any application for the third law. Its use is primarily in chemical thermodynamics, which is beyond the reach of an early introduction such as this.

3.8 Signals and noise

We showed earlier in section 3.5 that entropy is much more than a thermodynamic property and that in some sense it is concerned with information about probability distributions. We take this a little further now by considering the flow of entropy through a system, such as that shown in Fig. 3.8.1. The system outlined in this figure is of a channel along which energy is being transferred by a set of reversible engines, operating in the forward and reverse mode as shown. With this system there is a flow of energy from left to right, and it can be readily seen that the energy flow consists of a work rate $\dot{W}$ and a heat rate $\dot{Q}$. We assume that the maximum and minimum temperatures are T_S and T_R for each cycle, the working medium is a perfect gas and that the cycles of the engines are as shown in Fig. 3.8.2 so that $p_2 = p_4$.

Then we have from the diagram and equation [3.5.12]

$$s_3 - s_4 = s_2 - s_4$$

$$= c_v \ln \left(\frac{p_2}{p_4} \cdot \frac{v_2^\gamma}{v_4^\gamma} \right)$$

$$= c_v \ln \left(\frac{p_2}{p_4} \cdot \frac{T_2^\gamma}{T_4^\gamma} \cdot \frac{p_4^\gamma}{p_2^\gamma} \right)$$

$$= c_v \ln \left(\frac{p_4^{\gamma-1}}{p_2^{\gamma-1}} \cdot \frac{T_2^\gamma}{T_4^\gamma} \right) \qquad [3.8.1]$$

but $p_4 = p_2$ and therefore

$$s_3 - s_4 = \gamma c_v \ln \left(\frac{T_2}{T_4} \right)$$

$$= c_p \ln \left(\frac{T_S}{T_R} \right)$$

The rate at which entropy is transferred during the cycle is $C = \dot{M}(s_2 - s_4)$, where $\dot{M}$ is the mass flow rate of the working medium. The entropy transfer rate C is also the rate at which

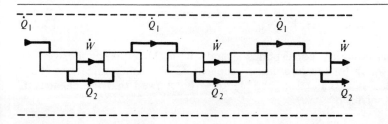

Fig. 3.8.1. Entropy flow system.

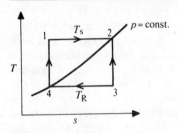

Fig. 3.8.2. Process representation.

entropy is transferred along the channel, and thus we have for the entropy transfer rate of the channel,

$$C = c_p \dot{M} \ln \left(\frac{T_S}{T_R} \right)$$ [3.8.2]

or since

$$\left(\frac{T_S}{T_R} \right) = \frac{\dot{Q}_S}{\dot{Q}_R} = \frac{(\dot{W} + \dot{Q}_R)}{\dot{Q}_R}$$

then this can be written

$$C = c_p \dot{M} \ln \left(\frac{\dot{W} + \dot{Q}_R}{\dot{Q}_R} \right)$$

$$= c_p \dot{M} \ln \left(1 + \frac{\dot{W}}{\dot{Q}_R} \right)$$ [3.8.3]

The ratio $(\dot{W}/\dot{Q}_R)$ is interesting since it represents, in effect, the ratio between the *mechanical* energy transfer $\dot{W}$ and the *thermal* transfer $\dot{Q}_R$ along the channel. The mechanical transfer $\dot{W}$ can be regarded as ordered energy flow, and $\dot{Q}_R$ disorganised energy flow. In other words the ratio $(\dot{W}/\dot{Q}_R)$ is the signal/noise ratio for the transfer process which, in the notation of communications theory, would be written $(\dot{W}/\dot{Q}_R) = (S/N)$. Then, in this notation equation [3.8.3] becomes

$$C = c_p \dot{M} \ln \left(1 + \frac{S}{N} \right)$$ [3.8.4]

This equation corresponds exactly with the equation derived by Shannon for the *information* capacity C of a communications channel. For a channel of bandwidth w he showed that the capacity C was given by

$$C = w \ln \left(1 + \frac{S}{N} \right)$$ [3.8.5]

in natural units, or correspondingly

$$C = w \log_2 \left(1 + \frac{S}{N}\right)$$ [3.8.6]

in the unit bit/sec. In either equation the ratio (S/N) is the signal/noise power ratio.

The similarity between equations [3.8.4] and [3.8.5] is not by chance and occurs because each system *is* transferring entropy or information. In the thermodynamic channel the rate of flow of information is the same function of signal/noise ratio as in the communications channel, and furthermore the product of the mass flow rate and specific heat is the equivalent of the bandwidth w in the communications channel.

It is worth noting that the thermodynamics result was derived from equation [3.8.1] by assuming that $p_2 = p_4$ and that normally we would have $p_2 > p_4$, in which case the channel capacity would, from inspection of equation [3.8.1], be less than this. It follows, then, that the channel capacity given by equation [3.8.4] is a *maximum* capacity – as is that given by the Shannon result.

The usual entropy units in thermodynamics and communications theory differ by a constant factor k, the Boltzmann constant of equation [3.5.2], and thus for the entropy S in communications theory we find in contrast to that equation

$$S = -\sum_i p_i \log_a p_i$$

where the logarithm base a is taken to be e for natural units and for bit information units $a = 2$. An interesting point is that a thermodynamic equation involving entropy can, if we wish, be written using the bit as the unit of information. As an example, if we take steam at temperature and pressure 250 °C and 500 kN/m^2, then the entropy $s = 7\cdot271$ kJ/kg K and thus

$$s = 7\cdot271 \text{ kJ/kg K}$$

$$= \frac{7\cdot271 \times 10^3}{1\cdot380 \times 10^{-23}} \quad \text{(natural information units/kg)}$$

$$= \frac{7\cdot271 \times 10^3}{1\cdot380 \times 10^{-23}} \times (\ln 2)^{-1} \quad \text{(bits/kg)}$$

$$= 7\cdot603 \times 10^{26} \quad \text{(bits/kg)}$$

The magnitude of this result shows that in thermodynamics and fluid mechanics we are dealing with an information flow rate many orders of magnitude higher than in any known communications system. A natural result since, in principal, we are dealing with phenomena which result from random molecular motion.

3.9 Road traffic systems

The flow of vehicles along a road, or a system of roads, can be analysed in a number of different ways. Many of these methods are statistical and we shall not attempt such an analysis here. Instead we shall confine ourselves to a preliminary study of dense traffic systems where the number of vehicles per unit length of road is high and it becomes more meaningful to talk of a density or concentration of vehicles on the road.

We define the *density* or *concentration* c, as being the number of cars per unit length of road, and the *flow*, q, as the number of vehicles that pass a line on the road in unit time. The principle of conservation of cars was previously discussed briefly in section 1.1 and we now apply equation [1.1.1] expressing the principle to a length of road δl over a period of time δt. In terms of c and q equation [1.1.1] becomes

$$\frac{\partial c}{\partial t} \cdot \delta l = -\frac{\partial q}{\partial l} \cdot \delta l \qquad [3.9.1]$$

and thus the equation of continuity for a traffic flow is

$$\frac{\partial c}{\partial t} + \frac{\partial q}{\partial l} = 0 \qquad [3.9.2]$$

On an actual road the flow q is a function of many parameters, but it is interesting to introduce a simplification at this point and assume that q is solely a function of the concentration c. This implies that the rate at which drivers pass along a road is only influenced by the vehicle density and is obviously only part of the story. However, if we make this assumption we have $q = q(c)$, and equation [3.9.2] may be written

$$\frac{\partial c}{\partial t} + \left(\frac{dq}{dc}\right) \cdot \frac{\partial c}{\partial l} = 0 \qquad [3.9.3]$$

If now dq/dc may be taken to be approximately a constant, say a_0, then we have for c the following differential equation

$$\frac{\partial c}{\partial t} + a_0 \frac{\partial c}{\partial l} = 0 \qquad [3.9.4]$$

which is satisfied by a solution of the form

$$c = f_1(l - a_0 t)$$

giving also

$$q = f_2(l - a_0 t)$$

The solution shows that variations in q and c travel along the road with an effective velocity of $a_0 = dq/dc$, or in other words as a *wave.* These waves are called *continuity waves* and may be observed in traffic flow before and after road obstructions such as traffic lights. If the mean velocity of the vehicles is v then we have, by analogy with equation [1.2.5], that $q = cv$ and thus

$$a_0 = \frac{dq}{dc} = v + c\frac{dv}{dc} \qquad [3.9.5]$$

Sensible drivers tend to reduce speed when the traffic density increases and only increase speed when the density reduces and thus dv/dc is normally negative, showing from equation [3.9.5] that a_0 can be either positive or negative, depending on the magnitude of c and dv/dc. Thus, when c is very large the changes in concentration travel backwards along the road, and when c is small the wave travels in the direction of the vehicle motion. A moving vehicle is always passing through continuity waves generated on the road ahead, and the greater the concentration ahead the greater is the approach velocity of the wave relative to the vehicle. It is for this reason that fog conditions require a reduction in mean vehicle speed to avoid collisions occurring because drivers are unable to react quickly to the information presented by the oncoming continuity wave. An ideal road is one in which the vehicle speed and continuity waves travel at the same velocity, so that the driver is never faced with the requirement for a quick reaction. Equation [3.9.5] shows that this can only be achieved if $c = 0$ – the road carries *no* vehicles.

The assumption made above that the traffic flow q is a function only of the concentration can be improved upon by noting that drivers are also influenced by the change in concentration ahead. For example, if the driver of a car sees that the vehicles in front of him are coming closer together he reacts by slowing down. Assuming this reaction is linear, then we obtain the improved relationship

$$q = q\left(c, \frac{\partial c}{\partial l}\right)$$

or

$$\frac{dq}{dl} = \frac{\partial q}{\partial c} \cdot \frac{\partial c}{\partial l} + \frac{\partial q}{\partial(\partial c/\partial l)} \cdot \frac{\partial^2 c}{\partial l^2}$$

Clearly $\partial q[\partial(\partial c/\partial l)]$ must be negative and we take it to have a constant value β, a parameter of the traffic system. Taking $\partial q/\partial c = a_0$ again we then have

$$\frac{\partial q}{\partial l} = a_0\frac{\partial c}{\partial l} - \beta\frac{\partial^2 c}{\partial l^2} \qquad [3.9.6]$$

Thus, combining equations [3.9.2] and [3.9.6], we get the new equation satisfied by c,

$$\frac{\partial c}{\partial t} + a_0 \frac{\partial c}{\partial l} - \beta \frac{\partial^2 c}{\partial l^2} = 0 \qquad [3.9.7]$$

The third term in this equation $\beta(\partial^2 c/\partial l^2)$ represents a diffusion process. In the case of the flow of traffic, it is this term that ensures that variations in the concentration along a road without traffic lights are eventually smoothed out. This is another example of the dissipative or smoothing effect of the second law. Again, it is the second law that insists that β is a positive quantity and fortunately so, for otherwise large-scale pile-ups of vehicles would be the inevitable result even on little-used roads in the countryside.

3.10 Mixing processes

In this section we are concerned with mixing processes that take place without chemical reaction between the various constituents involved in the mixture. We assume that the masses of the constituents can be independently varied, if required, and we use the symbol M_i for the mass of the ith constituent present in the mixture. The mass of the mixture is then M given by

$$M = \sum_i M_i$$

and we may now define a *mass fraction* for the ith component by the ratio (M_i/M). In section 2.5 it turned out to be useful to think in terms of the mole as the mass unit, and this again will be found to be the case here. Thus if n_i is the number of moles of constituent i then we define the *mole fraction* χ_i by the relationship

$$\chi_i = \frac{n_i}{\sum_i n_i}$$

$$= \frac{n_i}{N} \qquad [3.10.1]$$

For a non-reacting mixture with no significant electromagnetic energy contributions the two-property rule for a single substance (equation [3.5.3]) is insufficient since the entropy S will be a function of the relative proportions of the constituents as well as the total volume and internal energy. Thus, replacing equation [3.5.3], we have for the mixture

$$S = S(U, V, n_1, n_2, \ldots) \qquad [3.10.2]$$

The arguments given previously in section 3.5 to show how the absolute temperature and pressure can be determined in terms of partial derivatives of S still apply and hence we have

$$\left(\frac{\partial S}{\partial U}\right)_{V,n_1,n_2\ldots} = \frac{1}{T} \qquad [3.10.3]$$

and

$$\left(\frac{\partial S}{\partial V}\right)_{U,n_1,n_2\ldots} = \frac{p}{T} \qquad [3.10.4]$$

but now we have the additional possibility that partial derivatives of the type $(\partial S/\partial n_1)_{U,V,n_2,n_3}\ldots$ may also be of physical significance. We note that in the case of $(\partial S/\partial U)_{V,n_1,n_2}\ldots$ and $(\partial S/\partial V)_{U,n_1,n_2}\ldots$ the partial derivatives were shown to be equal to $(1/T)$ and (p/T), each having as denominator the absolute temperature T. We assume that in the case of the partial derivatives with respect to n_i this is again going to be the case, and therefore tentatively write

$$\left.\begin{aligned}\left(\frac{\partial S}{\partial n_1}\right)_{U,V,n_2,n_3\ldots} &= -\frac{\mu_1}{T} \\ \left(\frac{\partial S}{\partial n_2}\right)_{U,V,n_1,n_3\ldots} &= -\frac{\mu_2}{T}\end{aligned}\right| \qquad [3.10.5]$$

etc.

where μ_i is known as the *chemical potential*, and the negative sign in equation [3.10.5] is introduced to accord with convention. With equation [3.10.5] the Gibbs equation, previously equation [3.5.9] for the single substance, may now be deduced by differentiation of equation [3.10.2] and is

$$\delta s = \frac{\delta u}{T} + \frac{p}{T}\,\delta v - \frac{1}{T}\sum_i \mu_i\,\delta n_i$$

or

$$T\,\delta s = \delta u + p\,\delta v - \sum_i \mu_i\,\delta n_i \qquad [3.10.6]$$

If we consider the two interacting systems A and B, we showed earlier in section 3.5 that if A and B are simple substances, the condition for thermal and mechanical equilibrium is that equations [3.5.7] are both satisfied, i.e.

$$\left(\frac{\partial S_A}{\partial U_A}\right)_{V_A} = \left(\frac{\partial S_B}{\partial U_B}\right)_{V_B}$$

and

$$\left(\frac{\partial S_A}{\partial V_A}\right)_{U_A} = \left(\frac{\partial S_B}{\partial V_B}\right)_{U_B}$$

Only if these conditions are satisfied will the entropy be a maximum and the systems in thermal and mechanical equilibrium.

If now systems A and B are both mixtures, or substances which can mix, it is possible for diffusion of matter to take place across the boundary between A and B and hence for the molar masses n_{i_A} and n_{i_B} to vary. The condition that S is a maximum in this case thus has additional conditions which, by analogy with the development in section 3.5, may be deduced to be

$$\left(\frac{\partial S_A}{\partial n_{1_A}}\right)_{U_A, V_A, n_{2A}, n_{3A}, \dots} = \left(\frac{\partial S_B}{\partial n_{1_B}}\right)_{U_B, V_B, n_{2B}, n_{3B}, \dots}$$

or

$$\left(\frac{\mu_i}{T}\right)_A = \left(\frac{\mu_i}{T}\right)_B \qquad [3.10.7]$$

for the general case.

However, the two systems A and B were already in thermal equilibrium and thus $T_A = T_B$. The condition for equilibrium thus becomes simply

$$\mu_{i_A} = \mu_{i_B} \qquad [3.10.8]$$

Consider now what would be the result of A and B being at the same temperature and pressure with $\mu_{i_A} = \mu_{i_B}$ for all values of i except $i = j$, then we have

$$(\delta S_A + \delta S_B) = \frac{1}{T}(\mu_{j_B} - \mu_{j_A})\, \delta n_{j_A}$$

where δn_{j_A} is the net change in the molar mass of component j in A and is clearly equal to $-\delta n_{j_B}$. If the systems are isolated from their surroundings, then the second law requires that $(\delta S_A + \delta S_B) \geq 0$ and thus for δn_{jA} to be positive, i.e. mass transfer $B \to A$ then $\mu_{jB} \geq \mu_{jA}$. Thus the mass transfer will be in the direction of *falling* chemical potential μ_i.

We know from experience, that in the saturated region, a vapour and liquid can coexist at the same temperature and pressure with the relative proportions by mass remaining constant. The two phases are thus in equilibrium, and we then deduce that the two phases must have the same chemical potential.

3.11 Problems

1 A chemical plant produces a liquid petroleum product which has a density of 700 kg/m^3 and which has to be stored in a large tank at a pressure of 1300 kN/m^2. When the liquid is required for use it is passed through a valve after which its pressure is at

atmospheric pressure, $101 \, \text{kN/m}^2$. Find the power output that might be achieved if the valve could be replaced by a turbine, assuming an average flow rate from the tank of $20 \, \text{m}^3/\text{hr}$. Make reasonable estimates of the value of the energy that might be produced in this way and try to find out whether the cost of the turbine would ever be recovered.

2 The hydraulic braking system of a car contains a bubble of air of volume $2 \, \text{cm}^3$ when the oil is at atmospheric pressure. When the brakes are applied the oil pressure rises to a value of $300 \, \text{kN/m}^2$ above that of the atmosphere. Assuming that the compression process on the bubble takes place at constant temperature, find the reduction in volume of the bubble and the work done by the driver that is not usefully applied to the operation of the brakes. In an emergency application of the brakes the compression of the air is so rapid that the heat transfer to the oil is negligible. Estimate the work done in this case.

3 Seamed stainless steel is produced from rolled strip by a continuous arc-welding process that takes place in an atmosphere of argon. The stainless steel enters the machine at the rate of $20 \, \text{kg/hr}$ and at a room temperature of $300 \, \text{K}$. At the outlet the tube has an average temperature of $315 \, \text{K}$ and the flow of argon passes to waste down its centre at sensibly the same temperature.

   If the welding machine consumes $5 \, \text{kg/hr}$ of argon and is thermally insulated on the outside, estimate the electrical power consumption of the arc welder.

   Assume that for argon $c_p = 0\cdot525 \, \text{kJ/kg K}$, and for steel $c_p = 0\cdot51 \, \text{kJ/kg K}$.

4 A ramjet engine takes in air at $100 \, \text{kN/m}^2$, $288 \, \text{K}$ and having a velocity of $200 \, \text{m/sec}$. At the exhaust nozzle the pressure is again $100 \, \text{kN/m}^2$, but the temperature and velocity are increased to $1200 \, \text{K}$ and $600 \, \text{m/sec}$. Assuming that the fluid at inlet and outlet has the properties of air and neglecting any increase in mass flow due to the fuel, find the heat input per unit mass flow.

5 In a steam turbine working at inlet conditions of $15\,000 \, \text{kN/m}^2$, $800 \, \text{K}$, a steam bleed of 10% of the inlet mass flow is taken from a mid-stage of the turbine at conditions of $3000 \, \text{kN/m}^2$, $600 \, \text{K}$. The steam from this bleed is then passed through a smaller turbine operating a compressor plant. The exhaust from both turbines is arranged to be at the same condition of $5 \, \text{kN/m}^2$ pressure and $0\cdot95$ dryness fraction. If the inlet mass flow to the main turbine is $200 \, \text{kg/sec}$ find the power developed by the main and auxiliary turbines.

6 A computer module for an artificial satellite is contained in a box which is filled with nitrogen at a pressure of $150 \, \text{kN/m}^2$. After launch the box develops a leak through a small hole approximately $2 \, \text{mm}$ in diameter. Estimate the rate at which the

nitrogen escapes through the hole, making whatever assumptions seem reasonable. The satellite is not pressurised and for nitrogen $\gamma = 1\cdot4$.

7 An aircraft navigation light is positioned in the leading edge of the wing tip. The aircraft capability at sea-level is Mach $0\cdot5$ and it is restricted to an altitude of 15 000 m where the pressure and temperature are $12\cdot1 \, kN/m^2$ and $216\cdot7$ K respectively. Find values for the maximum and minimum pressure and temperature that the landing light should be designed to withstand.

8 A geothermal power station is to use a steam supply generated from volcanic activity and the maximum available temperature is of the order of 800 K. A nearby river, to be used for cooling water supply, has a temperature variation from 280 K to 300 K throughout the year. In order to preserve marine life the maximum permissible heating of the river must not exceed 50 MW in summer and 60 MW in winter. Estimate the maximum possible values for the electrical power output of the station during summer and winter conditions. Explain what factors make the result an estimate rather than a precise calculation.

9 Give reasons why in a spacecraft requiring electrical power a preferred solution is to use an array of solar panels generating electrical power directly, rather than to have some form of heat engine on board with a fuel and oxygen supply.

10 A test report on a refrigerator compressor gives the following information:

Working fluid	Freon 12
Inlet temperature	223 K
Inlet pressure	$123\cdot7 \, kN/m^2$
Outlet temperature	298 K
Outlet pressure	$415 \, kN/m^2$

Use thermodynamic tables for Freon 12 to establish whether these results are credible or not.

11 A steam turbine has an isentropic efficiency of 90% and at inlet the steam conditions are $700 \, kN/m^2$, 573 K. The outlet pressure is $50 \, kN/m^2$. Find the outlet condition of the steam, the change in entropy and the loss of availability. Assume that the flow through the turbine is adiabatic.

Chapter 4

Energy conversion

4.1 Introduction

Little of our modern way of life would exist without energy-conversion processes. We perhaps take for granted water, gas and electricity supplies in our homes and offices. Road, rail and air transport are also seen as basic necessities. All of these daily requirements depend upon the conversion of energy from óne form to another, and, as we have seen in our earlier work, it is rarely possible to do this without at the same time producing some proportion of the energy in a comparatively useless or invaluable form.

The object of this chapter is to look in more depth at energy conversion and related processes, and where possible to demonstrate ways in which a particular conversion process can be improved by optimum *economic* design. Many practical conversion processes are far too complex to be dealt with here, but the basic optimisation processes are little different for a modern power station and a lawn-mower engine.

4.2 Thermodynamic cycles

In Chapter 3 we introduced the Carnot cycle as representing an ideal reversible cycle which could be used to obtain a work output equal to the difference between the energy supplied by heating and that rejected by cooling. A cycle cannot be constructed in which there is no heat-rejection process, however desirable this may seem. Unfortunately no practical engine can be frictionless, and hence in practice we are unable to construct the ideal reversible engine. A consequence of this is that a practical engine can only be made to operate approximately to the Carnot cycle and will inevitably have a conversion ratio or thermodynamic efficiency less than that of the ideal Carnot cycle engine. In practice, it may thus prove possible to obtain higher conversion ratios by abandoning the Carnot cycle altogether and replacing it by alternative cycles. In examining alternative power-plant cycles, it is quite insufficient to make comparisons using the conversion ratio parameter by itself. Comparison must take into account the economic consequences of building, maintaining and running the plant required to operate to the thermodynamic cycle. The object of this section is to try to make sensible comparisons between power plants operating to various cycles and to point out some of the economic factors that are involved in power-plant design.

Unlike the ideal Carnot cycle, it is necessary to consider the working medium when examining practical power plants, and in particular whether it is to be a vapour power plant or one using a gas, for example air. The difference is important, because each medium requires quite different plant and equipment for the various processes in the cycle and this in turn will affect the economics of the plant. We consider first the steam or vapour power plant.

Steam or vapour power plant

In Fig. 4.2.1 the ideal Carnot cycle is shown on a temperature-entropy plot or Ts diagram. The power plant required to take a vapour medium round the cycle is shown to the left of the Ts diagram. The Carnot cycle as shown consists of the following processes:

(1–2) Heat supply at constant temperature.
(2–3) Isentropic expansion with work output.
(3–4) Heat rejection at constant temperature.
(4–1) Isentropic compression with work input.

Since the cycle is depicted in Fig. 4.2.1 to lie completely within

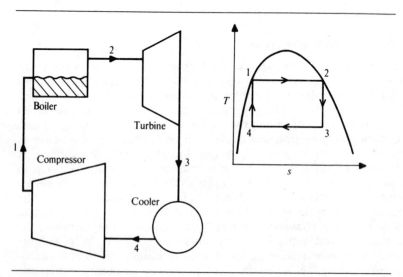

Fig. 4.2.1. The Carnot cycle.

the wet region, it follows from our previous discussions that processes (1–2) and (3–4) – as well as being at constant temperature – are also constant-pressure processes.

There are a number of reasons why the Carnot cycle is not used in practical power plants. Perhaps the major reason is that the compression process (4–1) ends with the working medium in the liquid state, and this presents severe design and operating problems. A second substantial difficulty is that of controlling the heat-rejection process (3–4) so that the medium at state 4 is at the required temperature.

In the Carnot cycle the specific volume of the working fluid is high during the compression process, and as a result the compressor has a size comparable with that of the turbine; similar comparisons can be made of the costs involved. It turns out that the Carnot cycle is not an economic proposition for a vapour power plant, but it can be made into a practical cycle by a small modification. The modified cycle has the compressor replaced by a pump, and in order to do this the heat-rejection process is allowed to proceed until the medium is completely condensed. The resulting cycle is known as the *Rankine cycle*. The Rankine cycle, or ideal saturated steam cycle, together with a schematic diagram of the plant layout is shown in Fig. 4.2.2. In the *ideal* cycle the processes are all assumed to be frictionless but, as the *Ts* diagram shows, we have introduced a heating process (5–1) which is not now at constant temperature. The conversion ratio will therefore be less than that of the Carnot cycle. The Rankine cycle is comprised of the following processes.

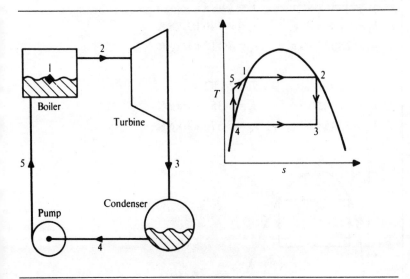

Fig. 4.2.2. The Rankine cycle.

(1–2) Heat supply at constant temperature and pressure.

(2–3) Isentropic expansion with work output.

(3–4) Heat rejection at constant temperature and pressure carried out until condensation is complete.

(4–5) Isentropic pumping process to raise the liquid pressure to that of the boiler.

(5–1) Heat supply at constant pressure – but not constant temperature – carried out until all the liquid is in a saturated state.

Example 4.2.1

Make a comparison between the Carnot and Rankine cycles for a steam power plant having maximum and minimum pressures of 3600 kN/m^2 and 3 kN/m^2 respectively.

Data

$p_1 = p_2 = p_5 = 3600 \text{ kN/m}^2$

$p_3 = p_4 = p_4' = 3 \text{ kN/m}^2$

(1–2–3–4–1) is the Carnot cycle

(1–2–3–4'–5–1) is the Rankine cycle

$s_2 = s_3, \qquad s_4 = s_1, \qquad s_4' = s_5$

Thermodynamic table values for saturated water and steam

At

$p = 3600 \text{ kN/m}^2, \qquad T = 244° \text{ C}$

$h_l = 1058 \text{ kJ/kg}, \qquad s_l = 2\cdot740 \text{ kJ/kg K}$

$h_g = 2802 \text{ kJ/kg}, \qquad s_g = 6\cdot113 \text{ kJ/kg K}$

At

$p = 3 \text{ kN/m}^2, \qquad T = 24 °\text{C}$

$h_l = 101 \text{ kJ/kg}, \qquad s_l = 0\cdot354 \text{ kJ/kg K}$

$h_g = 2545 \text{ kJ/kg}, \qquad s_g = 8\cdot576 \text{ kJ/kg K}$

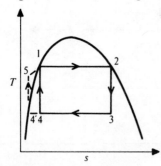

Cycle analysis and calculation

$s_3 = 0.354(1 - x_3) + 8.576 x_3 = s_2 = 6.113$

Therefore

$x_3 = 0.743$

$h_3 = 101(1 - 0.743) + 2545 \times 0.743$

$\quad = \underline{1917 \text{ kJ/kg}}$

$s_4 = 0.354(1 - x_4) + 8.576 x_4 = s_1 = 2.740$

Therefore

$x_4 = 0.290$

$h_4 = 101(1 - 0.290) + 2545 \times 0.290$

$\quad = \underline{810 \text{ kJ/kg}}$

For process $(4'-5)$

$\delta s = \delta u + p \, \delta v = 0$

Therefore

$\delta h = \delta u + p \, \delta v + v \, \delta p$

$\quad = v \, \delta p$

But v is sensibly constant for a liquid process, and hence

$$\int_{4'}^{5} dh = h_5 - h_{4'} = v(p_5 - p_{4'})$$

Therefore

$h_5 = 101 + 10^{-3}(3600 - 3)$

$\quad = \underline{105 \text{ kJ/kg}}$

CARNOT CYCLE

$W_{23} = h_2 - h_3 = 885 \text{ kJ/kg}$

$W_{41} = h_4 - h_1 = -248 \text{ kJ/kg}$

$Q_{12} = h_2 - h_1 = 1744 \text{ kJ/kg}$

$Q_{34} = h_4 - h_3 = -1107 \text{ kJ/kg}$

$\rho = -\dfrac{Q_{34}}{Q_{12}} = \mathbf{0.635}$

$\eta_{\text{Th}} = (1 - \rho) = \mathbf{0.365}$

Power output per unit steam flow $= (W_{23} + W_{41}) = W$

$W = \mathbf{637 \text{ kW/kg}}$

RANKINE CYCLE

$W_{23} = h_2 - h_3 = 885 \text{ kJ/kg}$

$W_{4'5} = h_{4'} - h_5 = -4 \text{ kJ/kg}$

$Q_{52} = h_2 - h_5 = 2697 \text{ kJ/kg}$

$Q_{34'} = h_{4'} - h_3 = -1816 \text{ kJ/kg}$

$\rho = -\dfrac{Q_{34'}}{Q_{12}} = \mathbf{0.673}$

$\eta_{\text{Th}} = (1 - \rho) = \mathbf{0.327}$

Power output per unit steam flow $= (W_{23} + W_{4'5}) = W$

$W = \mathbf{881 \text{ kW/kg}}$

The above example illustrates two important points. In the first place the Rankine cycle for the given condition has a conversion ratio that is 12% worse than that of the Carnot cycle, and this naturally would imply a 12% worse fuel consumption for a given power output. The second most relevant point is that, on the contrary, the specific power output of the Rankine cycle is some 28% better than that of the Carnot cycle and that in turn means a reduction of size, and hence capital cost, of 28%. In addition there are reductions in capital and maintenance costs resulting from the use of a pump rather than a compressor, which again favour the Rankine cycle. The conclusion then is that – even for the ideal frictionless cycles compared here – there may well be economic advantage in selecting a cycle with a *lower* conversion ratio.

Example 4.2.2
Repeat the comparison of Example 4.2.1, using this time realistic component efficiencies for the turbine compressor and pump. In the case of the turbine assume an isentropic efficiency of 0·9 and a corresponding value for the compressor and pump of 0·8.

Data

$$\eta_{23} = 0·9$$

$$\eta_{41} = \eta_{4'5} = 0·8$$

All other values as in Example 4.2.1.

Analysis and calculation

$$W_{23} = h_2 - h_3 = 0·9(W_{23})_{is}$$
$$= 0·9 \times 885$$
$$= 797 \text{ kJ/kg}$$

Hence

$$h_3 = \underline{2005 \text{ kJ/kg}}$$

$$W_{41} = h_4 - h_1 = \frac{(W_{41})_{is}}{0·8}$$

$$= -\frac{248}{0·8}$$

$$= -310 \text{ kJ/kg}$$

Hence

$$h_4 = \underline{748 \text{ kJ/kg}}$$

$$W_{4'5} = h_{4'} - h_5 = \frac{W_{4'5}}{0\cdot8}$$

$$= \frac{-4}{0\cdot8}$$

$$= -5 \text{ kJ/kg}$$

Hence

$$h_5 = \underline{106 \text{ kJ/kg}}$$

CARNOT CYCLE	RANKINE CYCLE
$Q_{12} = h_2 - h_1 = 1744 \text{ kJ/kg}$	$Q_{52} = h_2 - h_5 = 2696 \text{ kJ/kg}$
$Q_{34} = h_4 - h_3 = -1257 \text{ kJ/kg}$	$Q_{34'} = h_{4'} - h_3 = -1904 \text{ kJ/kg}$
$\eta_{Th} = \mathbf{0\cdot279}$	$\eta_{Th} = \mathbf{0\cdot293}$
Specific power output	Specific power output
$= W = (W_{23} + W_{41})$	$= W = (W_{23} + W_{4'5})$
$W = \mathbf{487 \text{ kW/kg}}$	$W = \mathbf{792 \text{ kW/kg}}$

The above example shows clearly that, with the component efficiencies selected, the power plant working to the modified Carnot cycle now has no advantage whatsoever, and any economic analysis must favour the plant based on the Rankine cycle. Conversion ratio is now 5% better for the Rankine plant and the specific power advantage has increased to 63% – a dominant economic pointer.

Further economic advantage can be obtained by modifying the cycle so that the maximum temperature in the cycle is increased. An obvious possible way is to increase the temperatures T_1 and T_2 in the cycle shown in Fig. 4.2.2. However, there is a preferred method which is to employ the superheat cycle illustrated in Fig. 4.2.3. In this cycle there is an additional heating process (2–2′), usually at constant pressure, which has two advantages. One is that the conversion ratio is improved by the higher maximum temperature, and the second is that it is now possible to have an expansion process (2′–3′) which is outside the saturation curve. This enables turbine design and maintenance to be simplified, and in that respect reduces costs.

The superheat cycle of Fig. 4.2.3 is but a small stage in the process of steam cycle improvement and much more is done, particularly in plants for producing electricity or marine propulsion. The thermodynamic analysis of the more complex cycles is similar to that already given here and can be found in many texts dealing with steam-plant design. Our interest must now turn from

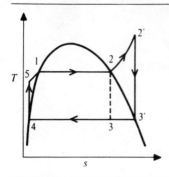

Fig. 4.2.3. The modified Rankine cycle.

vapour power plants to the corresponding development of the cycle analysis of gas power plants.

A cycle based on the Carnot cycle is not a practical possibility for gas power plants because of the near impossibility of economically achieving heat supply and rejection at constant temperature. An ideal gas power cycle is illustrated in Fig. 4.2.4 and is known as the Brayton cycle. The characteristic processes of the ideal Brayton cycle are as follows:

(1–2) Isentropic compression with work input.
(2–3) Heat supply at constant pressure.
(3–4) Isentropic expansion with work output.
(4–1) Heat rejection at constant pressure.

The schematic diagram of the plant in Fig. 4.2.4 shows a *closed* plant in which the working media is continually recycled through the components of the plant. However, since the working medium is often air, advantage can be taken of this by using the earth's atmosphere as the heat-rejection component. In such an *open*-plant process (4–1) takes place in the atmosphere, inlet to the compressor being from the atmosphere and turbine exhaust to the atmosphere. The open plant also has the advantage that heat supply can take the form of combustion within the working medium, since effectively the exhaust products are not returned to the intake.

We now consider the analysis of the ideal Brayton cycle, using a perfect gas as the working fluid. With this assumption it is possible to investigate the cycle using analytical methods, rather than the computational approach that has to be adopted for vapour cycles.

In the cycle of Fig. 4.2.4, the processes (1–2) and (3–4) are isentropic and thus,

$$T_2 = T_1 \left(\frac{p_2}{p_1}\right)^{(\gamma-1)/\gamma}, \qquad T_3 = T_4 \left(\frac{p_3}{p_4}\right)^{(\gamma-1)/\gamma}$$

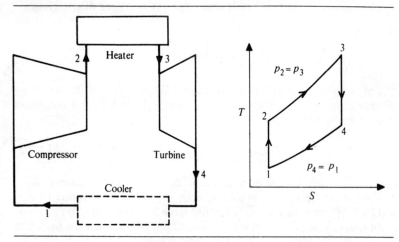

Fig. 4.2.4. The Brayton cycle.

but $p_2 = p_3$ and $p_4 = p_1$ and hence

$$\left(\frac{p_2}{p_1}\right) = \left(\frac{p_3}{p_4}\right) = r_p = \text{cycle pressure ratio}$$

Therefore

$$T_2 = T_1 r_p^{(\gamma-1)/\gamma}, \qquad T_3 = T_4 r_p^{(\gamma-1)/\gamma}$$
$$W_{12} = h_1 - h_2 = c_p(T_1 - T_2)$$
$$W_{34} = h_3 - h_4 = c_p(T_3 - T_4)$$
$$Q_{23} = h_3 - h_2 = c_p(T_3 - T_2)$$
$$Q_{41} = h_1 - h_4 = c_p(T_1 - T_4)$$

The conversion ratio η_{Th} is given by

$$\eta_{\text{Th}} = \left(1 - \frac{Q_R}{Q_S}\right) = \left(1 + \frac{Q_{41}}{Q_{23}}\right)$$

$$= 1 + \left(\frac{T_1 - T_4}{T_3 - T_2}\right)$$

$$= 1 - \left(\frac{1}{r_p}\right)^{(\gamma-1)/\gamma} \qquad\qquad [4.2.1]$$

Thus, for the ideal or frictionless Brayton cycle the conversion ratio is a function solely of the cycle pressure ratio r_p, and in fact η_{Th} increases monotonically with r_p.

The power output per unit mass flow $W = (W_{12} + W_{34})$, giving

$$W = c_p(T_1 - T_2) + c_p(T_3 - T_4)$$

$$= c_p T_1 (1 - r_p^{(\gamma-1)/\gamma}) + c_p T_3 \left(1 - \left(\frac{1}{r_p}\right)^{(\gamma-1)/\gamma}\right)$$

For a maximum value of W then $\partial W/\partial r_p = 0$, and hence if we assume fixed upper and lower temperatures T_3 and T_1 this condition gives a value for r_p of

$$r_p = \left(\frac{T_3}{T_1}\right)^{\gamma/\{2(\gamma-1)\}} \qquad [4.2.2]$$

which can readily be shown to occur at a maximum value of W.

Care must be taken, however, with the use of equations [4.2.1] and [4.2.2] since they are results that have been obtained by analysis of an ideal cycle, and in practice component efficiency must be taken into account. However, an interesting result is obtained if we take the value for r_p that gives a maximum specific work output and substitute it for r_p in equation [4.2.1]. We find for maximum W that

$$\eta_{Th} = 1 - \left(\frac{1}{r_p}\right)^{(\gamma-1)/\gamma}$$

which gives

$$\eta_{Th} = 1 - \left(\frac{T_1}{T_3}\right)^{1/2} \qquad [4.2.3]$$

With the same maximum and minimum values, the Carnot cycle gives a conversion ratio of

$$\eta_c = 1 - \frac{T_1}{T_3} \qquad [4.2.4]$$

and thus equation [4.2.3] shows that even the ideal cycle has a conversion ratio which can be considerably lower than that of the Carnot cycle at this condition of optimum W. For reasonable values say $T_1 = 300$ K, $T_3 = 1300$ K the reduction is of the order of 32%.

Example 4.2.3
A gas-turbine plant operating basically to a Brayton cycle has compressor and turbine isentropic efficiencies of 85% and 90% respectively. The inlet temperature to the compressor is 288 K and the cycle pressure ratio is 9·5. Compare the cycle parameters for two alternative maximum temperatures of 1000 K and 1300 K and assume the working medium is air with $\gamma = 1·4$ and $c_p = 1·005$ kJ/kg K.

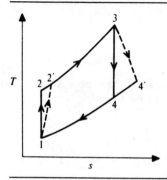

Data

$$\eta_c = 0.85$$
$$\eta_e = 0.9$$
$$T_1 = 288 \text{ K}$$
$$r_p = \frac{p_2}{p_1} = \frac{p_3}{p_4} = 9.5$$

(a) $T_3 = 1000$ K,   (b) $T_3 = 1300$ K

Cycle analysis
On the Ts diagram the cycle (1–2–3–4–1) has isentropic expansion and compression processes and cycle (1–2′–3–4′–1) is the actual cycle to which the plant is operating. We have

$$\eta_c = \frac{h_2 - h_1}{h_{2'} - h_1} = \frac{c_p(T_2 - T_1)}{c_p(T_{2'} - T_1)}$$

Therefore

$$T_{2'} = T_1 + \frac{T_2 - T_1}{\eta_c}$$

But

$$\frac{T_2}{T_1} = \left(\frac{p_2}{p_1}\right)^{(\gamma-1)/\gamma}$$

and hence we can find $T_{2'}$.

For the expansion process (3–4′) we have correspondingly

$$\eta_e = \frac{h_3 - h_{4'}}{h_3 - h_4} = \frac{c_p(T_3 - T_{4'})}{c_p(T_3 - T_4)}$$

giving

$$T_{4'} = T_3 - \eta_e(T_3 - T_4)$$

However,

$$\frac{T_3}{T_4} = \left(\frac{p_3}{p_4}\right)^{(\gamma-1)/\gamma}$$

and hence $T_{4'}$ is found.

The temperatures T_1, $T_{2'}$, T_3 and $T_{4'}$ are now known and enable the conversion ratio, etc. to be found.

Calculation

$$T_2 = T_1\left(\frac{p_2}{p_1}\right)^{(\gamma-1)/\gamma} = 288 \times 9 \cdot 5^{0 \cdot 4/1 \cdot 4}$$

$$= 548 \text{ K}$$

$$T_{2'} = 288 + \frac{548 - 288}{0 \cdot 85}$$

$$= 594 \text{ K}$$

$$T_4 = T_3\left(\frac{p_4}{p_3}\right)^{(\gamma-1)/\gamma}$$

(a) $T_4 = 1000\left(\frac{1}{9 \cdot 5}\right)^{0 \cdot 4/1 \cdot 4}$ (b) $T_4 = 1300\left(\frac{1}{9 \cdot 5}\right)^{0 \cdot 4/1 \cdot 4}$

$\quad\quad = \underline{526 \text{ K}}$ $\quad\quad\quad\quad\quad\quad = \underline{683 \text{ K}}$

$T_{4'} = 1000 - 0 \cdot 9(1000 - 526)$ $\quad\quad T_{4'} = 1300 - 0 \cdot 9(1300 - 683)$

$\quad\quad = \underline{573 \text{ K}}$ $\quad\quad\quad\quad\quad\quad = \underline{745 \text{ K}}$

$W_{12'} = 1 \cdot 005(288 - 594) = -308 \text{ kJ/kg},$
$W_{34'} = 1 \cdot 005(1000 - 573) = 429 \text{ kJ/kg},$
$Q_{2'3} = 1 \cdot 005(1000 - 594) = 408 \text{ kJ/kg},$
$Q_{4'1} = 1 \cdot 005(288 - 573) = -286 \text{ kJ/kg},$

$\quad\quad\quad\quad W_{12'} = 1 \cdot 005(594 - 288) = -308 \text{ kJ/kg}$
$\quad\quad\quad\quad W_{34'} = 1 \cdot 005(1300 - 745) = 558 \text{ kJ/kg}$
$\quad\quad\quad\quad Q_{2'3} = 1 \cdot 005(1300 - 594) = 710 \text{ kJ/kg}$
$\quad\quad\quad\quad Q_{4'1} = 1 \cdot 005(288 - 745) = -429 \text{ kJ/kg}$

$W = W_{12'} + W_{34'} = 429 - 308 = \textbf{121 kJ/kg}$

$\quad\quad\quad\quad W = W_{12'} + W_{34'} = 558 - 308 = \textbf{250 kJ/kg}$

$$\eta_{\text{Th}} = \left(1 + \frac{Q_{4'1}}{Q_{2'3}}\right) = \left(1 - \frac{286}{408}\right)$$

$$= \textbf{0·299}$$

$$\eta_{\text{Th}} = \left(1 + \frac{Q_{4'1}}{Q_{2'3}}\right) = \left(1 - \frac{429}{710}\right)$$

$$= \textbf{0·396}$$

The corresponding conversion ratios for a Carnot cycle with the same T_1 and T_3 are

$$\eta_c = \left(1 - \frac{288}{1000}\right), \qquad \eta_c = \left(1 - \frac{288}{1300}\right)$$

$$= 0\cdot712 \qquad\qquad = 0\cdot778$$

and are obviously considerably higher than those achieved in either practical cycle.

The effect of increasing temperature on the specific power output and conversion ratio of the actual cycle is quite marked and indicates the advantage to be gained from improved materials which allow increased turbine entry temperatures. An alternative method to obtain an improved conversion ratio is to develop the cycle itself by introducing additional compression, expansion and thermal processes at appropriate sections of the cycle. Generally, such modifications have the effect of introducing additional complexity in the plant itself and often an increase in weight, thus making them quite unsuitable for aircraft usage. For land-based plant it is common to find more complex cycles than the simple Brayton cycle discussed here.

Another common gas power cycle is that upon which the petrol engine depends, called the *Otto* cycle. The details of the ideal Otto cycle are given in Fig. 4.2.5 on both a *Ts* and *pv* diagram. The Otto cycle consists of four processes as follows:

(1–2) Isentropic compression with work input.
(2–3) Heat intake at constant volume.
(3–4) Isentropic expansion with work output.
(4–1) Heat rejection at constant volume.

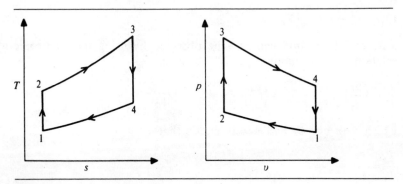

Fig. 4.2.5. The Otto cycle.

As in the case of the Brayton cycle the actual plant for a practical cycle may employ heat rejection in the atmosphere, with combustion as the heat-intake process. In the practical plant, expansion and compression is normally done with a piston and cylinder arrangement, and thus equation [3.3.2] must be used for obtaining the work input or output for the processes (1–2) and (3–4). We have from this equation

$$\Delta W = p \, \delta v - \Delta \Phi$$

but since we are here considering adiabatic frictionless processes $\Delta \Phi = 0$ and thus

$$\Delta W = p \, \delta v$$
$$= T \, \delta s - \delta u$$
$$= -\delta u$$

Consequently, we have for the processes (1–2) and (3–4) that

$$W_{12} = u_1 - u_2 = c_v(T_1 - T_2)$$

and

$$W_{34} = u_3 - u_4 = c_v(T_3 - T_4)$$

For the heating and cooling processes we use the equation of thermal interaction [1.5.15] again with $\Delta \Phi = 0$, giving

$$\Delta Q = \delta u + p \, \delta v$$
$$= \delta u$$

Hence

$$Q_{23} = u_3 - u_2 = c_v(T_3 - T_2)$$

and

$$Q_{41} = u_1 - u_4 = c_v(T_1 - T_4)$$

Further information can now be obtained by using the isentropic relationships of equation [2.2.20] to give

$$\frac{T_2}{T_1} = \left(\frac{v_1}{v_2}\right)^{\gamma-1}, \qquad \frac{T_3}{T_4} = \left(\frac{v_4}{v_3}\right)^{\gamma-1}$$

which with $v_1 = v_4$ and $v_2 = v_3$ shows that

$$\frac{T_4}{T_1} = \frac{T_3}{T_2}$$

The conversion ratio η_{Th} can now be deduced, for we have

$$\eta_{Th} = 1 - \frac{Q_R}{Q_S} = 1 + \frac{Q_{41}}{Q_{23}}$$

$$= 1 - \left(\frac{T_4 - T_1}{T_3 - T_2}\right)$$

$$= 1 - \frac{T_1}{T_2}\left\{\frac{(T_4/T_1) - 1}{(T_3/T_2) - 1}\right\}$$

$$= 1 - \frac{T_1}{T_2}$$

But

$$\frac{T_1}{T_2} = \left(\frac{v_2}{v_1}\right)^{\gamma-1} = \left(\frac{1}{r_v}\right)^{\gamma-1}$$

where $r_v = (v_1/v_2)$ is known as the *compression ratio*. Hence we have

$$\eta_{Th} = 1 - \left(\frac{1}{r_v}\right)^{\gamma-1} \qquad [4.2.5]$$

which shows clearly that, for the ideal cycle, the conversion ratio increases monotonically with increasing compression ratio.

The specific work output $W = (W_{12} + W_{34})$ and thus

$$W = c_v(T_1 - T_2) + c_v(T_3 - T_4)$$

$$= c_v T_1(1 - r_v^{\gamma-1}) + c_v T_3\left(1 - \left(\frac{1}{r_v}\right)^{\gamma-1}\right)$$

Again, we may find a maximum value if we assume that the lowest and highest temperatures T_1 and T_3 are fixed. In fact the maximum specific work output can be shown to occur when r_v satisfies the following relationship:

$$r_v = \left(\frac{T_3}{T_1}\right)^{1/\{2(\gamma-1)\}} \qquad [4.2.6]$$

and consequently, at values of r_v higher than this, there is something of a dilemma presented in that the conversion ratio η_{Th} increases with r_v while the specific work output W decreases. The size of machine is roughly proportional to W, and requirements here contrast with the demand for a reduced fuel consumption or high value for η_{Th}. In practical designs the cycle does not in any way approach that of the ideal Otto cycle, and consequently further simplified analysis could be misleading.

4.3 Electrical conversion

There are no significant sources of energy in the form of electrical potential energy in the world, and as a result our electrical power requirements have to be obtained by an energy conversion from our natural energy resources of hydrocarbons, uranium and the potential or kinetic energy of water and air resulting from meteorological phenomena. These last two resources have an advantage where electrical conversion is concerned in that the conversion process is quite independent of any limitations resulting from the Carnot theorem. No heating process is involved and the only restriction on the conversion efficiency is that brought about by dissipation – mainly as a result of frictional dissipation. Conversion from a thermal source has, in addition to frictional dissipation, irreversibilities resulting from heat-transfer processes, and perhaps even more significant the overriding bound to the conversion efficiency resulting from the Carnot theorem.

Whatever the energy source it is conventional in large-scale electrical generation to first convert the energy supply to a mechanical work output and then use this as a work input to an electrical generator. Any dissipation effects within the mechanical/electrical conversion process are consequently similar in the two cases. It is interesting, therefore, to compare the dissipation and other limitations in the two types of process producing a work and electrical output. The relevant equations illustrating the point are [1.4.15] for the frictional dissipation, [3.4.10] for the Carnot limitation and [2.6.4] for the electrical dissipation. We have, therefore, the most important effects in Table 4.3.1.

All present-day methods of large-scale electrical power generation suffer from the dissipative effects or limitations listed in the table, but for small-scale generation it is a practical proposition to consider alternative methods, including what has become known as

Table 4.3.1

Energy source	Work output limiting equations	Mechanical/electrical conversion dissipation
Meteorological phenomena	$\Delta W = -\delta\left(\dfrac{v^2}{2}\right) - \dfrac{\delta p}{\rho} - \delta\phi - \Delta\Phi$ for each flow process	$\dot{\Phi}_e = \rho\, J^2$ for each electric current flow
Combustion or nuclear fission	$\eta_{Th} = \dfrac{W}{Q_S} \le \left(1 - \dfrac{T_R}{T_S}\right)$ for each conversion process and $\Delta W = -\delta\left(\dfrac{v^2}{2}\right) - \dfrac{\delta p}{\rho} - \delta\phi - \Delta\Phi$ for each flow process	$\dot{\Phi}_e = \rho\, J^2$ for each electric current flow

direct energy conversion whereby chemical energy within a fuel – or a thermal source of energy – is used directly to produce an electrical power output. There is no intermediate stage in these processes and consequently frictional dissipation due to fluid motion is largely insignificant. If the energy source is thermal, then the device is still subject to the overall limitation imposed by the Carnot theorem, but if the process is one of chemical reaction within the device and the energy conversion takes place as a result of this, then the Carnot theorem has no relevance. Thus, the process of energy conversion within an electric battery is *not* subject to any restriction imposed by the Carnot theorem.

The fuel cell is an example of a conversion device in which chemical energy is converted directly into an electrical power output on a continuous basis, rather than as a charge and discharge process as in an electrical battery. A sketch of a typical hydrogen/oxygen fuel-cell configuration is shown in Fig. 4.3.1. In the cell the anode and cathode are made of a conducting material that is sufficiently porous to allow oxygen and hydrogen to diffuse under pressure through the material, and between the anode and cathode is an electrolyte – usually potassium hydroxide. The electrolyte is important to the reaction, and its purpose is to allow diffusion of ions between the porous conductors. When an electrical load is applied between the anode and cathode the reactions taking place are as follows. At the anode, oxygen passing through to the surface takes electrons and reacts with the water to form an OH^{-ve} hydroxyl ion. The ions then diffuse through the electrolyte to the cathode, where a

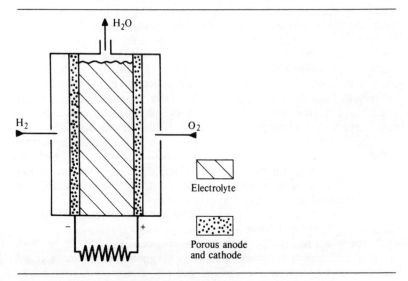

Fig. 4.3.1. Schematic drawing of a hydrogen/oxygen fuel cell.

reaction takes place with the hydrogen forming water, in the process releasing an electron which then passes round the external circuit. Thus the process, simplified in this description, allows hydrogen and oxygen to react together producing water, and in the process transfers electrons round the external circuit. No temperature difference is required and there is thus no limit to the conversion ratio imposed by the Carnot theorem. However, there are many other practical problems and dissipations that provide practical limits to the conversion efficiency.

The example of a fuel cell shown in Fig. 4.3.1 uses hydrogen and oxygen and is perhaps the easiest to construct and operate. The only complications found are that for satisfactory operation the porous conductors have to embody a catalyst for the reaction to proceed at a satisfactory rate, and that the electrolyte becomes diluted and later requires replacement. Fuel cells can be constructed to use other fuels, but in many cases the reaction only takes place at a high temperature, and furthermore unless the fuels are highly purified the cell may become contaminated and the reaction rate decrease. At present the practical application of fuel cells is confined to systems in space vehicles where the practical utility of obtaining an electrical power output, without the problems associated with a thermal power plant, make it an attractive proposition. It is interesting to observe that all human beings, and indeed all living systems, use a direct conversion process to obtain an electrical power output from oxidation of a fuel or food.

As an alternative to a direct conversion process, we may consider some new method of generating electricity directly from a thermal source which would allow the use of higher operating temperatures than can be tolerated in conventional power plants using turbomachinery. Although such a plant cannot have a conversion ratio greater than that of the reversible engine operating over the same temperature range, the advantages of being able to increase the maximum temperature significantly are worth pursuing. One type of plant being developed to operate at greatly increased temperatures makes use of the well-known electrical phenomena that when a conductor is moved through a magnetic field it has an e.m.f. induced in it, and if the circuit is complete then a current flows. Normal generation plant makes use of this by mechanically moving metallic conductors through the field, but in principle the metallic conductors can be replaced by an ionised gas. This is the principle behind the magnetohydrodynamic conversion process.

In Fig. 4.3.2 the basic ideas of this method of generation are shown in a simplified form.

In the diagram, ionised gas enters the conversion duct at high velocity and across the duct is a strong magnetic field. The charged particles moving within the ionised gas will, if unconstrained, move at right angles to both the direction of their motion and the

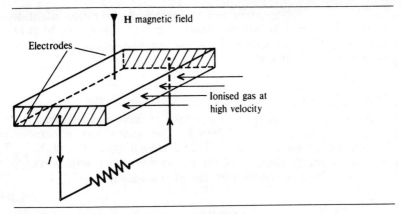

Fig. 4.3.2. Magnetohydrodynamic conversion process.

magnetic field. However, the ionised gas is constrained to remain within the duct, and therefore it induces an electric field between the two conducting electrodes which in turn produces a current in the external load. The electrical power output is the result of a decrease in the kinetic energy of the ionised gas, and the energy of the working fluid is reduced at outlet from the conversion duct. The application of these simple principles produces immense practical problems requiring further development at present. A simplified version of the type of plant being considered for large-scale generation is shown in Fig. 4.3.3. Here it can be seen that the M.H.D.

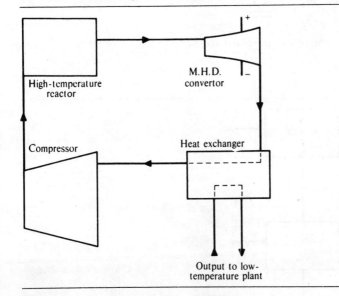

Fig. 4.3.3. A possible scheme for an M.H.D. power plant.

plant has a complementary low-temperature generation plant to ensure minimum loss of availability, and to allow the M.H.D. converter to operate within a narrow but high-temperature band where it is most efficient.

Example 4.3.1
An M.H.D. power plant is expected to have a maximum operating temperature of 4000 K. The plant rejects energy to a conventional power plant at a temperature of 1700 K. Make a rough estimate of the proportion of the total work output which is supplied by the M.H.D. plant.

Data
$T_1 = 4000$ K, $T_2 = 1700$ K
$T_3 = 300$ K (estimate).

Analysis
Make an estimate of $\dot{W}_1/(\dot{W}_1 + \dot{W}_2)$ on the basis of the Carnot plant. Then we have, for the M.H.D. plant,

$$\frac{\dot{Q}_1}{\dot{Q}_2} = \frac{T_1}{T_2}$$

and hence

$$\frac{\dot{Q}_1 - \dot{Q}_2}{\dot{Q}_2} = \frac{\dot{W}_1}{\dot{Q}_2} = \frac{T_1 - T_2}{T_2}$$

For the conventional plant we have

$$\frac{\dot{Q}_2}{\dot{Q}_3} = \frac{T_2}{T_3}$$

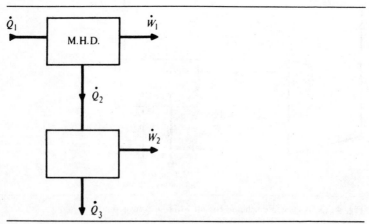

and thus

$$\frac{\dot{Q}_2}{\dot{Q}_2 - \dot{Q}_3} = \frac{\dot{Q}_2}{\dot{W}_2} = \frac{T_2}{T_2 - T_3}$$

Comparing the two results we can then find the ratio $\dot{W}_1/(\dot{W}_1 + \dot{W}_2)$.

Calculation

$$\frac{\dot{W}_1}{\dot{Q}_2} = \frac{4000 - 1700}{1700} = 1 \cdot 353$$

$$\frac{\dot{Q}_2}{\dot{W}_2} = \frac{1700}{1700 - 300} = 1 \cdot 214$$

$$\frac{\dot{W}_1}{\dot{W}_2} = 1 \cdot 353 \times 1 \cdot 214 = 1 \cdot 643$$

$$\frac{\dot{W}_1}{\dot{W}_1 + \dot{W}_2} = \frac{1 \cdot 643}{1 \cdot 643 + 1}$$

$$= \mathbf{0 \cdot 622}$$

4.4 Nuclear energy

In our previous work we have assumed that the conservation principles for mass and energy were valid, and in normal experimentation we would not be able to detect any error which cast doubt on these assumptions. However, it was first postulated by Einstein that energy and mass are equivalent, and that when energy is released in a chemical reaction the total mass of the constituents is reduced. The equation relating the energy release and mass decrease is now well known and is

$$E = Mc^2 \tag{4.4.1}$$

where E is the energy release, M is the reduction in mass and c is the velocity of light in a vacuum. Inserting a value of $c = 2 \cdot 998 \times 10^8$ m/sec into equation [4.4.1] gives

$$E = M(2 \cdot 998 \times 10^8)^2 \text{ J}$$

$$= 8 \cdot 988\, M \times 10^{13} \text{ kJ}$$

Thus, for example, when 1 kg of a hydrocarbon fuel is burnt with oxygen, giving an energy release of the order of 45×10^3 kJ the total reduction in mass of the fuel and oxygen atoms is M, given by

$$M = \frac{45 \times 10^3}{8 \cdot 988 \times 10^{13}}$$

$$= 5 \times 10^{-10} \text{ kg}$$

The reduction in mass is clearly quite undetectable under normal circumstances, and indicates that we shall not be in serious error if we retain the principle of conservation of mass for the processes we have discussed so far.

Normal chemical reactions take place because of processes involving the electrons in the outer shell of the atom and leave the nucleus of the atom quite unchanged. The nucleus containing protons and neutrons accounts for by far the largest contribution to the mass of the atom, and it follows that if the mass of the nucleus could be changed this could be equivalent to a substantial quantity of energy. Some basic definitions are now required in order to take this discussion a little further.

An atom which has no net electric charge is said to be neutral and has the same number of electrons in the electron shells as protons in the nucleus. An electron and proton have equal and opposite electric charge, but the mass of the proton is 1835 times that of the electron. The number of protons is called the *atomic number Z* and identifies the *element*, i.e. mercury, oxygen, etc. Within the atomic nucleus of a particular element there may be a number of uncharged particles called *neutrons*, each having a mass slightly more than the proton and 1837 times that of the electron. The number of neutrons does not change the chemical characteristics of the element, but the mass of the atom changes significantly. The various forms of the element, having different numbers of neutrons within the nucleus, are called *isotopes* and are distinguished by the number of neutrons N. By convention ^{16}O, ^{17}O and ^{18}O are the usual symbols for three isotopes of the element oxygen, and in each case the superscript indicates the total number of neutrons in the atomic nucleus. The total number of protons and neutrons is called the *mass number A*, which is an integer approximation to the atomic weight of the element isotope. Thus we have

Z = Number of protons, identifies the element and is the atomic number in the periodic table.

N = Number of neutrons, identifies the isotope of the element.

A = mass number and is the integer nearest to the atomic weight.

Thus the isotope ^{235}U of uranium, atomic number $Z = 92$, has $(235 - 92)$ neutrons in the atomic nucleus. Similarly, the most common isotope of Oxygen ^{16}O atomic number 8, has a total of 8 neutrons in the nucleus. We can now calculate the mass of the elemental particles making up this atom, if we assume the electron mass $m_e = 9 \cdot 108 \times 10^{-31}$ kg. We have for the ^{16}O atom

8 protons: mass $= 8 \times 1835 \times m_e$
8 neutrons: mass $= 8 \times 1837 \times m_e$
8 electrons: mass $= 8 \times m_e$

$$\text{Total} = 2 \cdot 676 \times 10^{-26} \text{ kg}$$

However, measurements made on the isotope ^{16}O show that the actual atomic mass is in fact $2 \cdot 656 \times 10^{-26}$ kg, a mass defect of 2×10^{-28} kg per atom which is the mass equivalent of the binding energy or energy required to assemble the atom from the individual protons, neutrons and electrons. For 1 kg of ^{16}O the total binding energy within the atoms is $6 \cdot 7 \times 10^{11}$ kJ – an enormous quantity of energy. Unfortunately, it is not possible at present to release this energy by any practical process suitable for power production, but it is possible to obtain a small proportion of the binding energies of certain elements by means of the process called *nuclear fission*.

A simplified diagram of the nuclear-fission process is shown in Fig. 4.4.1. In the figure N is the nucleus of an atom being bombarded by a neutron. If the neutron has the correct velocity and the atom has been suitably chosen, it is possible for the collision to break the nucleus apart and the elementary particles then re-form again in what can be different elements or isotopes. In some cases the splitting and re-forming process takes place several times until a stable outcome is achieved, and in doing so more neutrons and other elementary particles may be released as shown in the diagram. In some cases the nuclear reaction can require a net energy input, and in other cases the process gives a net energy output in the form of kinetic energy of the particles or internal energy of the molecules. However, even the simplest nuclear reaction is exceedingly complex in detail, and even with the simplified picture shown in Fig. 4.4.1 it can be seen that an interesting possibility presents itself if the reaction process produces more neutrons than are required to trigger the process initially. If this is the case, then it is likely that a chain reaction called *fission* can be produced, as shown in Fig. 4.4.2, providing that the quantity of material present is sufficient to allow the reaction to proceed, even with the inevitable loss of neutrons due to flux through the boundary surface.

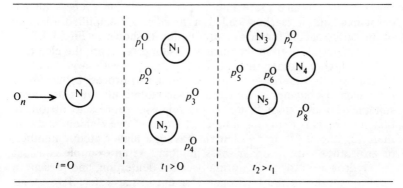

Fig. 4.4.1. The nuclear fission process.

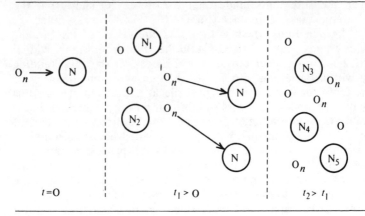

Fig. 4.4.2. A nuclear chain reaction.

The process depicted in Fig. 4.4.2 is called a *chain reaction* and can release enormous quantities of energy in a short time, as in an atomic bomb, or – if the process is slowed down by interspersing another element that reduces the neutron flux called a moderator – the reaction can be controlled and the energy released at a rate suitable for power production. This, in essence, is what takes place in the reactor of a nuclear power plant. To start the reaction it is necessary to have a material that emits neutrons naturally, and the most suitable natural material for this is the Uranium isotope ^{235}U. When the reaction is taking place the neutrons can be used to bombard other materials that are not in themselves suitable for the process, but which will break apart to produce other elements and isotopes suitable for fission. Thus we find the isotope ^{238}U is used within the reactor, because under neutron bombardment it pro- duces, among other things, Plutonium, a man-made element that can then be used in a subsequent fission reaction. In some cases the reactor can produce more material for the fission process than it consumes and it is then called a *breeder*. A simplified diagram of an early reactor for power production is shown in Fig. 4.4.3.

The materials used in the nuclear-fission process are the elements with the higher atomic numbers having isotopes with large numbers of neutrons, but there is an alternative nuclear process called *fusion* for which the elements with the lower atomic numbers are more suitable. To an extent fusion is the opposite process to fission, in that the object is to take a number of atoms of one element and fuse them together to form an element having a higher atomic number, or an isotope with a higher mass number. As an example consider the Hydrogen isotope ^2H, normally called deuterium, and present in small quantities in the Hydrogen element in sea-water. Two atoms of ^2H contain two protons, two neutrons and two electrons which

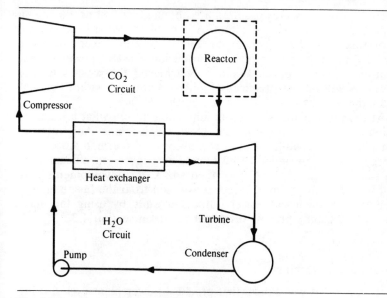

Fig. 4.4.3. An early nuclear power plant.

are exactly the same numbers as are found for the Helium isotope ^{4}He of atomic number 4. Thus, in principle the two atoms of ^{2}H can produce one atom of ^{4}He if the fusion process takes place, but it is by no means obvious whether this will require an energy input or produce an energy output. The matter can be resolved only by comparing the actual atomic masses of the two isotopes and thereby finding the change in binding energy for the process. Using tabulated values for the masses we have

$$\text{Actual mass of } ^2\text{H atom} = 3{\cdot}344 \times 10^{-27} \text{ kg}$$
$$\text{Actual mass of two } ^2\text{H atoms} = 6{\cdot}688 \times 10^{-27} \text{ kg}$$
$$\text{Actual mass of } ^4\text{He atom} = 6{\cdot}646 \times 10^{-27} \text{ kg}$$
$$\text{Difference in mass} = 4{\cdot}2 \times 10^{-29} \text{ kg}$$

Therefore energy release from fusion of 1 kg ^{2}H is

$$Mc^2 = \frac{4{\cdot}2 \times 10^{-29}}{6{\cdot}688 \times 10^{-27}} \times 8{\cdot}988 \times 10^{13}$$
$$= 5{\cdot}64 \times 10^{11} \text{ kJ}$$

This figure for the energy release is quite staggering and means, for example, that in a fusion process 30 kg of deuterium could produce an energy output of 500 MW for a complete year. Unfortunately, the fusion reaction will only take place at extreme temperatures such as those existing in the sun, and much development of methods of containing gases at these temperatures still requires to be done. The fusion reaction is achieved within the hydrogen bomb, where the temperatures required are obtained by first producing an explosive nuclear fission reaction. A controlled fusion process on a laboratory scale has been made, but it may be several decades before a practical power plant is producing a useful power output from the fusion process. When it is achieved the world's energy resources will lie – in practically limitless amounts – within the sea-water that covers four-fifths of the earth's surface.

At the present time it is quite impossible to say what a practical fusion power plant will look like. However, it seems likely that the first plants to be built will use a conventional thermal process for energy conversion and thus be limited in conversion efficiency by the requirements of the Carnot theorem. Later developments may well include some form of direct conversion from the fusion energy supply to an electrical power output, possibly by using the magnetohydrodynamic process discussed briefly in section 4.3.

4.5 Rotating machines

In section 4.2 it was shown that practical power-plant cycles, using both vapour and gas fluid media, depend upon the use of work input and output devices to perform the expansion and compression processes. There are many ways in which such devices can be designed and the purpose of this section is to indicate the type of analysis that can be performed on one class of device, the rotating machine.

We consider first the energy balances within a rotating system such as a turbine impeller, or the rotating blade system of an axial compressor. In Fig. 4.5.1 we have a rectangular Cartesian coordinate frame OXYZ which is rotating about the OZ axis with constant angular velocity ω. A particle of mass M is at a fixed point P in the XY plane with the radial distance $OP = r$.

If the particle is at a fixed point in the rotating coordinate system,

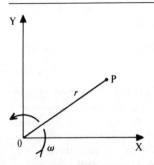

Fig. 4.5.1. Particle in a system of rotating axes.

then its angular velocity about the OZ axis is ω, and this gives rise to a radial force acting on the particle of magnitude $Mr\omega^2$ in the direction $\overrightarrow{PO}$. Suppose now that the particle is moved to some point P' where $OP' = (r + \delta r)$, then the element of work done by the particle in the repositioning is $Mr\omega^2 \delta r$ and is, in principle, recoverable if the repositioning is reversed. Thus to an observer, in the rotating axes the situation is exactly the same as if he were observing a particle in a potential energy field having a magnitude of $\phi = -(r^2\omega^2/2)$ per unit mass. For an elemental change in r of δr the work done per unit mass ΔW is given by

$$\Delta W = +\frac{d\phi}{dr}\,\delta r = -\omega^2 r\,\delta r \qquad [4.5.1]$$

which accordingly is what would normally be expected in a potential field. Let us now suppose that the particle is part of a fluid flow which is to be observed from the system of rotating axes, and in that system has velocity components (v'_r, v'_θ, v'_z) in the usual notation. If we ignore gravitational potential energy, the energy of the fluid per unit mass and the corresponding entrance and exit work $\delta\varepsilon$ is now given by

$$\delta\varepsilon = \delta M\left(u + pv + \frac{v'^2_r + v'^2_\theta + v'^2_z}{2} - \frac{r^2\omega^2}{2} \right) \qquad [4.5.2]$$

which, when compared with the corresponding result for the stationary axes (equation [1.5.6]) can be seen to be of the same form apart from the subtraction of the term $r^2\omega^2/2$. We can now use equation [4.5.2] to develop a steady-flow energy equation for flow observed in the rotating system.

Consider, then, some rotating part of a machine as shown in Fig. 4.5.2 with fluid entering at the cross-section (1) and leaving at (2). In the rotating-axis system, fixed relative to the rotating passage, the passage walls are stationary and hence the fluid flow does no work. Assuming the flow is adiabatic, the energy equation for the flow

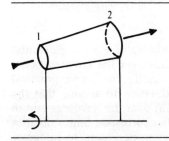

Fig. 4.5.2. Flow through a rotating machine.

between sections (1) and (2) is, from equation [4.5.2],

$$\left(h+\frac{v_r'^2+v_\theta'^2+v_3'^2}{2}-\frac{r^2\omega^2}{2}\right)_1=\left(h+\frac{v_r'^2+v_\theta'^2+v_3'^2}{2}-\frac{r^2\omega^2}{2}\right)_2 \qquad [4.5.3]$$

If now (v_r, v_θ, v_z) are the *absolute* velocity components of the flow relative to a set of *stationary* axes, then

$$v_r = v_r', \qquad v_\theta = (v_\theta' + \omega r) \quad \text{and} \quad v_z = v_z'$$

Substituting for v_r', v_θ' and v_z' in equation [4.5.3] we find that after rearranging, the energy equation becomes

$$\left(h+\frac{v_r^2+v_\theta^2+v_z^2}{2}-r\omega v_\theta\right)_1=\left(h+\frac{v_r^2+v_\theta^2+v_z^2}{2}-r\omega v_\theta\right)_2 \qquad [4.5.4]$$

In the stationary frame of reference, the normal steady-flow energy equation [1.5.11] applies, which for adiabatic flow may be written

$$\dot{W}=\dot{M}\left(h+\frac{v_r^2+v_\theta^2+v_z^2}{2}\right)_1-\dot{M}\left(h+\frac{v_r^2+v_\theta^2+v_z^2}{2}\right)_2 \qquad [4.5.5]$$

and here $\dot{W}$ is not necessarily zero. Combining equations [4.5.4] and [4.5.5] we find

$$\dot{W}=\dot{M}\omega(r_1 v_{\theta_1} - r_2 v_{\theta_2}) \qquad [4.5.6]$$

which gives the work done per unit mass, in the flow through the rotating passage, having angular velocity ω in the same sense as the velocity component v_θ of the absolute flow.

Equation [4.5.6] is the basis for analysis of the flow through rotating machinery. At no point in the derivation has it been assumed that frictional dissipation is negligible and the result is valid regardless of frictional dissipation. As written, equation [4.5.6] is in terms of the absolute velocity v_θ, but at times it is more convenient to use when expressed in terms of the relative velocity v_θ'. Substitution for v_θ yields

$$\dot{W}=\dot{M}\omega(r_1 v_{\theta_1}' - r_2 v_{\theta_2}')+\dot{M}\omega^2(r_1^2 - r_2^2) \qquad [4.5.7]$$

Equations [4.5.6] and [4.5.7] are strictly valid for an elemental stream tube where the difference in radius over the cross-sectional area is small compared to the radius itself. For many practical machines it may well be a valid approximation to assume that the work done per unit mass is nearly constant over the flow area and to take average values for (r_1, r_2) and the corresponding velocities $(v_{\theta_1}, v_{\theta_2})$. However, in some cases it will be necessary to integrate over the entrance and exit cross-sectional areas.

Example 4.5.1

A centrifugal water pump has an impeller which rotates at 500 r.p.m. Water enters the impeller through stationary vanes at a radius of 0·1 m and with an absolute velocity $v_{\theta_1} = 5$ m/sec, measured in the same sense as the impeller blade velocity. The impeller vanes are shaped so that the value of v_{θ_2} is 10 m/sec at an outlet radius of 0·25 m. If the mass flow rate of water is 7 kg/sec find the work required to drive the impeller, and the maximum pressure rise that could be produced by the pump, assuming that the inlet and outlet pipes have the same diameter.

Data

$r_1 = 0·1$ m, $r_2 = 0·25$ m

$v_{\theta_1} = 5$ m/sec, $v_{\theta_2} = 10$ m/sec

$\omega = \dfrac{2\pi \times 500}{60} = 52·4$ rad/sec

$\dot{M} = 7$ kg/sec

Analysis

From equation [4.5.6]

$$\dot{W} = \dot{M}\omega(r_1 v_{\theta_1} - r_2 v_{\theta_2})$$

giving $\dot{W}$.

If the inlet and outlet pipes are of the same diameter then, since $\rho = $ constant for water, the continuity equation [1.2.5] shows that the inlet and outlet velocities must be the same. Using the momentum equation [1.4.16] we have, for no change in potential energy,

$$\int_3^4 \frac{dp}{\rho} + \Phi_{34} + W_{34} = 0$$

where the inlet and outlet cross-sections are given the suffixes 3 and 4. For water the density is constant and hence

$$\int_3^4 \frac{dp}{\rho} = \frac{p_4 - p_3}{\rho} = -W_{34} - \Phi_{34}$$

We have that $\dot{W} = \dot{M}W_{34}$ is negative for a pump and hence the effect of the frictional dissipation – a positive quantity – is to reduce the pressure rise $(p_4 - p_3)$. Thus

$$(p_4 - p_3)_{max} = -\rho \frac{\dot{W}}{\dot{M}}$$

Calculation

$$\dot{W} = 7 \times 52 \cdot 4 (0 \cdot 1 \times 5 - 0 \cdot 25 \times 10)$$
$$= -734 \, \text{W}$$
$$= -0 \cdot 734 \, \text{kW}$$

The practical designs of rotating machines are diverse; however, the analysis of the thermodynamics and fluid mechanics of the fluid flow is in principle little different, regardless of the variation in layout and complexity. In principle the methods of analysis involve separating the device into a set of rotating and stationary passages and dealing with each separately. For flow within a stationary passage the normal equations for mass, momentum and energy balance apply, and within the rotating passage this is still the case, providing care is taken to include the potential energy term as in equation [4.5.3] and also velocity components in the *rotating* frame of reference. Equations [4.5.6] or [4.5.7] are used to give the work input or output to the fluid from the rotor vanes or blades.

4.6 Semiconductor devices

In section 2.6 we introduced the class of materials known as semiconductors which act rather differently in an electric field compared to the normal conducting materials such as copper. Semiconductor materials tend to have a high resistivity compared to conducting metals and the variation with temperature is much higher.

The method of charge transport in a semiconductor is rather different and takes place by two distinct processes. A thorough description of the charge transport requires the use of quantum mechanics which is quite outside the scope of this text. However, a brief discussion of the process will indicate the important features that mark the distinction between conductors and semiconductors.

We begin by considering an atom composed of a nucleus surrounded by electrons in a series of shells, each shell containing a number of electrons. The most strongly bound electrons in an atom are those in the inner shells close to the nucleus. The electron, or electrons, in the outer shell are called the *valence* electrons and it is these electrons that determine the properties of the atom for chemical reaction. There is a maximum possible number of electrons in each particular shell for a neutral atom, or an atom that is not ionised, and the elements in the periodic table are formed as the various possible shells, in effect, fill up with electrons from the nucleus outwards. The inert gases have complete shells; in contrast hydrogen and the alkali metals have only one electron in the outer shell, the remainder being complete. The halogens, on the other hand, have one electron less than the maximum in the outer shell and thus combine readily with hydrogen and the alkali metals to form a stable compound in which the valence electrons form a complete shell for one of the atomic members.

Electrical conduction in a crystalline solid depends upon the valence electrons being free to move. However, in the crystal lattice the separation distance between the atoms is of the same order of magnitude as the effective size of the atom. A consequence of this is that the valence electrons become shared between nearby atoms, rather than each belonging to a particular atom. In a conductor the possible energy levels of the shared electrons are such that some electrons are free to be accelerated by an electric field and then later recaptured by an atom or group of atoms in the lattice. In insulating materials the allowed energy levels forbid this process, and the conduction electrons available for the charge transport process are few.

In a semiconductor there are usually conduction electrons available for charge transport under an electric field, but there is also an additional process which does not exist in a metallic conductor. In effect the electrons available for charge transport, having left the atom to which they belonged, leave behind a *hole* in part of the valence shell which can later be filled by another electron. The result is that we have an additional process in which there is a drift of holes from atom to atom, which was shown by Dirac in 1932 to constitute a charge transport process. In a semiconductor, then, the two processes of charge transport in an electric field are (a) transport by the conduction electrons, and (b) transport due to the drift of holes in the valence band where an electron would normally be present. The drift of holes is equivalent to a flux of *positively* charged particles, i.e. of the opposite charge to the electron.

The ratio of the contributions to charge transport from the drift of electrons and holes can be varied by introducing small quantities of impurities during the growth of the semiconductor crystal. A semiconductor whose properties have been modified in this way is said to be *doped*. If the material is doped so that the predominant conduction mode is by electrons, or negative charges, the material is known as an N-type semiconductor, if by holes, then the material is known as a P-type semiconductor.

Let us consider the effect of operating the device shown in Fig. 4.6.1. When the switch is closed conduction and valence electrons in the material flow round the circuit. The electrons entering the P material in zone (2) combine with holes, and in doing so energy is released within the material. In contrast, at the junction in region (1) electrons leave the N material to pass through the junction and in doing so holes are produced in the N material, and energy supply being required to do this. As a consequence of these processes zone (1) becomes cooler and zone (2) hotter. For further operation on a

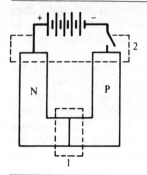

Fig. 4.6.1. A semiconductor device.

continuous basis the device then requires a thermal energy supply in zone (1) and gives thermal rejection in zone (2). The device operates as a reversed engine and can be used for refrigeration with zone (1) as the thermal absorption region. Reversal of the battery polarity reverses the heating and cooling effects in zones (1) and (2) and thus the device can be used for heating or cooling in zone (1) by simple operation of a switch.

The phenomena of heating or cooling at a junction when an electric current flows is known as the *Peltier effect,* and although the effect is more marked at a semiconductor junction it also occurs at the junction between two dissimilar metals. The magnitude of the effect for two materials A and B is assessed quantitatively by means of a *Peltier coefficient* $\pi(T)$, which is defined as the rate of energy output per unit current flow, i.e.

$$\pi_{AB}(T) = \frac{\dot{Q}_P}{I} \qquad [4.6.1]$$

where $\dot{Q}_P$ is the rate of energy output *due to the Peltier effect* and must be clearly distinguished from the *dissipation* $\dot{\Phi}_e$ of equation [2.6.4]. The parameter $\pi_{AB}(T)$ is a function of the temperature and is a joint property of the two materials A and B. It was first shown by Kelvin that the Peltier coefficient may be expressed in terms of absolute properties of the individual materials A and B by the equation

$$\pi_{AB}(T) = T(\varepsilon_B - \varepsilon_A) \qquad [4.6.2]$$

where ε_A and ε_B are known as the *absolute thermoelectric powers* of the materials A and B, and T is the absolute temperature. Again ε_A and ε_B are generally functions of temperature.

Combining equations [4.6.2] and [4.6.1] we find finally

$$\dot{Q}_P = IT(\varepsilon_B - \varepsilon_A) \qquad [4.6.3]$$

An important point made clear by equation [4.6.3] is that the Peltier effect is reversible in direction, as indicated earlier. We now con-sider the application of the Peltier effect to a thermoelectric re-frigerator shown in Fig. 4.6.2.

We let the thermal conductivity and resistivity of the two semiconductor materials be κ_N, κ_P and ρ_N, ρ_P respectively, the cross-sectional areas A and the distance between the junction and the current collectors l. If we assume that the only heat transfers to or from the system are $\dot{Q}_1$ at the cold zone and $\dot{Q}_2$ at the hot zone, then the energy balance for zone (1) gives

$$\dot{Q}_1 = \dot{Q}_P - A\left\{\kappa_N\left(\frac{dT}{dx}\right)_N + \kappa_P\left(\frac{dT}{dx}\right)_P\right\} \qquad [4.6.4]$$

where the bracketed term is the energy input to the junction by thermal conduction through the semiconductor materials.

Substituting for $\dot{Q}_P$ using equation [4.6.3] gives

$$\dot{Q}_1 = IT_1(\varepsilon_P - \varepsilon_N) - A\left\{\kappa_N\left(\frac{dT}{dx}\right)_N + \kappa_P\left(\frac{dT}{dx}\right)_P\right\} \qquad [4.6.5]$$

In order to proceed further we now require to know the tempera-ture gradients $(dT/dx)_N$ and $(dT/dx)_P$ at the junction surface in zone (1). In order to find these we consider a small element of one of the semiconductors, as shown in Fig. 4.6.3. The element is of volume $A\,\delta x$ and hence the actual rate of dissipation of electrical energy is, from equation [2.6.4],

$$\begin{aligned}
\dot{\Phi}_e &= \rho\frac{T^2}{A^2}(A\,\delta x) \\
&= \frac{\rho I^2\,\delta x}{A}
\end{aligned} \qquad [4.6.6]$$

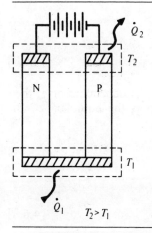

Fig. 4.6.2. A thermoelectric refrigerator.

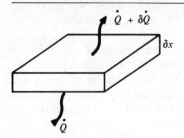

Fig. 4.6.3. Heat transfer through an element of semiconductor.

An energy balance on the system of Fig. 4.6.3 then gives

$$(\dot{Q} + \delta\dot{Q}) - \dot{Q} = \rho\frac{I^2\,\delta x}{A}$$

or

$$\frac{\mathrm{d}\dot{Q}}{\mathrm{d}x} = \frac{\rho I^2}{A}$$

but

$$\dot{Q} = -A\kappa\frac{\mathrm{d}T}{\mathrm{d}x}$$

and hence

$$\frac{\mathrm{d}\dot{Q}}{\mathrm{d}x} = -A\kappa\frac{\mathrm{d}^2T}{\mathrm{d}x^2}$$

giving finally

$$-\kappa\frac{\mathrm{d}^2T}{\mathrm{d}x^2} = \frac{\rho I^2}{A}$$

or

$$\frac{\mathrm{d}^2T}{\mathrm{d}x^2} = -\frac{\rho I^2}{\kappa A} \qquad\qquad [4.6.7]$$

If we may assume that ρ and κ are constant independent of T, then the solution to equation [4.6.7] that satisfies the boundary conditions ($x = 0$, $T = T_1$; $x = l$, $T = T_2$) is

$$T = -\frac{\rho I^2 x^2}{2\kappa A} + \left\{\frac{T_2 - T_1}{l} + \frac{\rho I^2 l}{2\kappa A}\right\}x + T_1$$

giving

$$\left(\frac{\mathrm{d}T}{\mathrm{d}x}\right)_{x=0} = \left\{\frac{T_2 - T_1}{l} + \frac{\rho I^2 l}{2\kappa A}\right\} \qquad\qquad [4.6.8]$$

Substitution for $(dT/dx)_{x=0}$ into equation [4.6.5] gives the result

$$\dot{Q}_1 = IT_1(\varepsilon_P - \varepsilon_N) - A\left\{\frac{(\kappa_N + \kappa_P)(T_2 - T_1)}{l} + \frac{\rho I^2 l}{A}\right\}\qquad[4.6.9]$$

For given values of current I and temperature T_1 the requirement that $\dot{Q}_1$ be as high as possible dictates that κ_N, κ_P and ρ be as low as possible. Clearly the semiconductor materials are likely to have an advantage over metals in this application.

Example 4.6.1
A thermoelectric refrigerator is constructed from semiconductor materials having the following properties: $\kappa_N = \kappa_P = 1\cdot5$ W/m K, $\rho_N = \rho_P = 1\cdot3 \times 10^{-1}$ ohm-m, $\varepsilon_P = -\varepsilon_N = 200 \times 10^{-6}$ V/K. Assume the length of each semiconductor element is 2×10^{-3} m and the cross-sectional area $0\cdot6 \times 10^{-6}$ m^2. Find the thermal input and output of the device for a current of 10 A and junction temperatures of 300 K and 450 K.

Data

$\kappa_N = \kappa_P = 1\cdot5$ W/m K
$\rho_N = \rho_P = 1\cdot3 \times 10^{-1}$ ohm-m
$\varepsilon_P = -\varepsilon_N = 200 \times 10^{-6}$ V/K
$l = 2 \times 10^{-3}$ m, $\qquad A = 0\cdot6 \times 10^{-6}$ m^2
$I = 10$ A
$T_1 = 300$ K, $\qquad T_2 = 450$ K

Analysis and calculation
Substituting values in equation [4.6.9] we find

$$\dot{Q}_1 = 10 \times 300 \times 400 \times 10^{-6} - 0\cdot6 \times 10^{-6}\left\{\frac{3 \times (450 - 300)}{2 \times 10^{-3}} + \right.$$
$$\left. + \frac{1\cdot3 \times 10^{-1} \times 10^2 \times 2 \times 10^{-3}}{0\cdot6 \times 10^{-6}}\right\}$$
$$= 1\cdot2 - 0\cdot16$$
$$= \mathbf{1\cdot04\ W}$$

The energy balance for $\dot{Q}_2$ is

$$\dot{Q}_2 = \dot{Q}_1 + \frac{I^2 A}{2\rho l}$$

Therefore

$$\dot{Q}_2 = 1\cdot04 + \frac{10^2 \times 0\cdot6 \times 10^{-6}}{2 \times 1\cdot3 \times 10^{-1} \times 2 \times 10^{-3}}$$
$$= \mathbf{1\cdot16\ W}$$

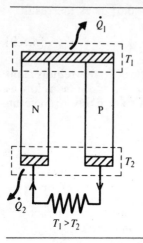

Fig. 4.6.4. Direct conversion in a semiconductor device.

The device shown in Fig. 4.6.2 can, if required, operate in a different mode if the battery is replaced by a resistor. In this mode (Fig. 4.6.4) the thermal input $\dot{Q}_1$ is less than the output $\dot{Q}_2$, zone (1) is at a higher temperature than zone (2) and the device in effect performs the function of a thermal power plant with direct conversion of thermal energy to electrical energy. If the resistance is removed, then an electric potential is generated between the two semiconductor elements, this phenomena being known as the *Seebeck effect*. A similar effect is found at the junction of two dissimilar metals and is put to practical use in the thermocouple.

4.7 Environmental considerations

We have seen in earlier sections that an inescapable consequence of all energy-conversion processes is an increase in entropy, and thereby a reduction in the available work output from the system and its environment. In order to use our knowledge of technology to produce the manufactured goods and services we require, it is essential that we use our available energy resources economically. Until comparatively recently this quest for economy was perhaps the main consideration when considering energy-conversion systems. However, over the years the realisation has come that environmental problems such as pollution, noise, reduction of amenities etc., are perhaps equally important and must be taken into account *in conjunction* with economic considerations. In this section we develop these ideas a little further by considering a number of particular examples where environmental considerations are important.

We begin by considering the effect, in different processes, of an energy input rate of 500 MW in a steady-flow process shown in Fig.

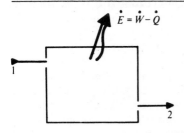

$\dot{E} = \dot{W} - \dot{Q}$

1

2

Fig. 4.7.1. The steady-flow system.

4.7.1. We give the data and results in tabular form and, since we are interested only in magnitudes, the calculations are rounded off to the first significant figure. In each case we are interested in the mass and volume flow rates required to give an energy transfer out of the system of $\dot{E} = 500$ MW a typical figure for the electrical output of a generating station.

The orders of magnitude given in the columns for $\dot{M}$ and $\dot{V}$ allow some interesting general observations to be made. We observe that in each process the changes in the variable parameters are typical of many found in natural phenomena – such as wind, running surface water, variations in atmospheric and sea temperature, etc. Consequently the orders of magnitude of $\dot{M}$ and $\dot{V}$ are the flow rates that would be involved in an energy-conversion process which resulted in 500 MW being obtained from the environment. Two observations can be made immediately. Firstly, as we look down the columns, both the mass and volume flow rates decrease spectacularly. Secondly, the order of reduction of the mass flow rate, 10^6, is small compared to the reduction in volume flow rate of 4×10^8.

It follows from this that if we are to extract energy from the environment processes near the top of the list we will require comparatively large power plants in order to achieve the required mass and volume flow rates. For example process (a) would need a windmill type of plant having a flow area of the order of 10^6 m^2 or one square kilometre, clearly a severe environmental problem compared to the fuel line of example (f) which requires a diameter of the order of 50 mm.

Process	Fluid	Variables	$\dot{M}$(kg/sec)	$\dot{V}$(m^3/sec)
(a) Decrease in k.e.	Air	$v_2 = 0, v_1 = 10$ m/sec	1×10^7	8×10^6
(b) Decrease in k.e.	Water	$v_2 = 0, v_1 = 10$ m/sec	1×10^7	1×10^4
(c) Decrease in T	Air	$(T_1 - T_2) = 10$ K	5×10^4	4×10^3
(d) Decrease in T	Water	$(T_1 - T_2) = 10$ K	1×10^4	1×10
(e) Change of phase	Water	$h_1 = h_g, h_2 = h_f$	2×10^2	$V_f = 2 \times 10^2$
(f) Combustion	Fuel oil (45 000 kJ/kg)	$T_1 = T_2, v_1 = v_2$	1×10	2×10^{-2}

The tabulated values also indicate that certain processes are quite unsuitable for the transmission of energy from one area of the country to another, again because of the high volume rates required, particularly in processes a, b and c. Furthermore, our knowledge of the Carnot theorem and its consequences leaves us in no doubt that if the required output is a work output, then transferring energy by means of small temperature differences necessarily produces a low conversion ratio which is quite uneconomical. The two processes that are highlighted by the table are those of the change of phase and combustion of fuel. Each of these has much reduced volume flow rates and consequently leads to economy in process plant construction and operation. Unfortunately, there are few natural sources of steam at high temperatures suitable for an energy-conversion process resulting in a work output, and consequently the majority of our power-producing plants employ combustion processes to produce the high-temperature steam. Steam is the working fluid in the majority of large power-producing plants and, as the table shows, this has advantages compared to a gas through the reduced volumetric flows.

Two alternative energy sources which have not been discussed in the table are the sun and nuclear fission, and it is interesting to see the orders of magnitudes involved. For solar energy the rate at which energy is received by the earth's surface, including the atmosphere, is of the order of $180\,MW/km^2$, and thus if this could be utilised the required output of 500 MW would need a surface area of around $3\,km^2$. Conversion processes using solar energy are in a rapid state of development at present and in the future may be a significant energy source. An output of 500 MW from a nuclear-fission process requires a consumption of fissile material of only $10^{-5}\,kg$ which, in orders of magnitude, is less than any values in processes (a),(b),(c),(d),(e) or (f). However, this one figure clouds the issue to some extent, since a complex process is required to produce the fissile material in a form suitable for a practical power plant.

The world energy crisis has concentrated attention on energy resources previously regarded as of little utility. As an example wave power from the sea is now regarded optimistically as a source of electrical energy for countries bordering the world's oceans. In other areas where climatic conditions are favourable, wind power can be shown to be feasible and an economic proposition.

4.8 Problems

1. A steam power plant operating to the Rankine cycle has maximum and minimum pressures of $3000\,kN/m^2$ and $0 \cdot 5\,kN/m^2$ and gives a net power output of 40 MW. The isentropic efficiency of the turbine is 87% and that of the pump 77%. Find the steam consumption and the heat intake in the boiler.

2. A gas turbine for a railway locomotive is designed to operate to a modified Brayton cycle in which, after compression, the air is first passed at constant pressure through a heat exchanger before combustion takes place. The heat exchanger is connected to the exhaust system of an auxiliary diesel engine and the rate of heat transfer between the two systems is 50 kW.

   In the gas turbine the maximum temperature is 1150 K, the air inlet temperature is 290 K, the mass flow is 5·0 kg/sec and the overall pressure ratio is 8·0. If the isentropic efficiencies of the turbine and compressor are 90% and 85% respectively, calculate the combustion heat intake and power output of the system. Assume the working medium is air and neglect the increased mass flow due to the fuel.

3. Using the data and diagram of Example 4.3.1 estimate the overall improvement in conversion ratio that has been achieved by having the heat input at 4000 K to the M.H.D. plant, rather than having the heat input to a conventional power plant at the lower temperature of 1700 K.

   Does this seem to be a worthwhile economic proposal?

4. A nuclear reactor of the breeder type would seen to produce more fuel than it consumes. Investigate this reactor system to determine how this can be done without contradicting the laws of thermodynamics.

5. The isotope lithium 7 has an atomic mass equal to $12·7796 \times 10^3$ times that of the electron mass. The atom has 3 protons, 4 neutrons and 3 electrons. Calculate the mass defect and binding energy for 1 kg of the isotope.

6. The rotor of an electrical alternator is cooled by hydrogen which enters the rotor at negligible whirl velocity. The cooling flow leaves the rotor at a radius of 0·4 m having passed through the internal passages in the machine. If the rotational speed of the alternator is 3000 r.p.m. and the mass flow rate of coolant is 3 kg/sec find the work required to be done on the hydrogen, assuming no change in pressure between inlet and outlet and neglecting frictional dissipation.

Chapter 5

Economics and energy conservation

5.1 Introduction

Until quite recently the main concern in energy transfer and conversion processes was one of overall economics, and little mention was made of the effect of processes upon long-term energy resources. It is now widely realised that the present forms of energy resource such as oil and coal are limited and indeed may only last over one life span at the current rate of depletion. There is little doubt that over this period of time alternative energy sources will be able to be utilised. The fusion process will enable hydrogen in sea-water to be used as a virtually limitless energy supply, or perhaps more effective methods of using solar energy may be developed.

At present these breakthrough developments are not available, and we must acquire a prudent attitude to the energy resources that we have and the methods by which we use them. It has become quite improper to investigate the economics of the process without at the same time taking into account the long-term effects of what is being done. To an extent, the long-term effects are already being taken into account by increased prices for fuel, reflecting the long-term scarcity of the raw material. This fiscal process is likely to carry on and perhaps even accelerate over the next decade, thus ensuring that our energy resources are used wisely and less wastefully.

It is the object in this chapter to look at energy transfer and conversion processes from an economic point of view, and to look into particular examples of systems design that assist energy conservation.

5.2 Efficiency and the use of fuel

Fuel resources in the earth are largely hydrocarbons existing in either a gas, liquid or solid form. When reaction with oxygen takes place at a high temperature, chemical energy is released from the fuel and the process is known as *combustion*. The chemical energy within the fuel is in effect stored solar energy, which was originally received by life on the surface of the earth many millions of years ago.

Knowledge of the chemical composition of the fuel enables the available energy release for a particular fuel to be found and the exhaust products, CO, CO_2, etc. determined. Unfortunately, practical combustion processes rarely conform to the ideal and some of the available energy is retained as chemical energy within the exhaust products. In most combustion systems the temperature of the exhaust products is much greater than that of the fuel and air prior to combustion, and thus the exhaust can transfer some of its energy by heating to another fluid. In gas-turbine systems it is the exhaust itself which is used as the working fluid for the turbine, but in most other cases a second fluid is used as the working medium in the subsequent processes.

Figure 5.2.1 shows a schematic diagram of a simple steam power plant.

The complete plant has three energy outputs; (a) a power output $\dot{W}$ from the turbine which is its useful purpose; (b) a heating to the condenser cooling fluid $\dot{Q}_R$ which normally produces a low value or even an undesirable heating effect in a river or the atmosphere; (c) an energy transfer in the exhaust in the form of chemical, internal or kinetic energy. In sections 3.4 and 4.2 it was shown that a thermal power plant must have heat rejection, although careful design of the plant and cycle could reduce its amount. What is now clear is that the heat-intake process to the plant may also be a source of energy rejection, through incomplete combustion or other effects within the boiler system.

We now consider the overall economics and profitability of the plant shown in Fig. 5.2.1 and make the assumptions given in the following table.

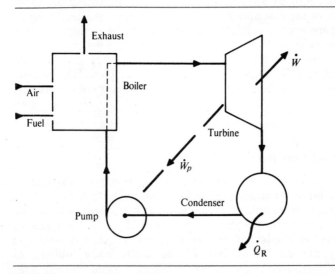

Fig. 5.2.1. A steam power plant.

Quantity	Symbol	Cost rate	Value rate	Comments
Work output rate	$\dot{W}$		$\alpha\dot{W}$	
Heat rejection	$\dot{Q}_R$		$\beta\dot{Q}_R$	Assumed that this has a positive value for heating, etc.
Energy supply rate available in fuel	$\dot{Q}_F$	$\gamma\dot{Q}_F$		
Energy rejection in exhaust	$\dot{Q}_E$		$-\delta\dot{Q}_E$	Assumed that pollution control laws make exhaust emission a liability
Operating and amortisation cost		$\varepsilon\dot{Q}_F$		

An energy balance gives for the system

$$\dot{Q}_F = \dot{Q}_E + \dot{Q}_R + \dot{W} \qquad [5.2.1]$$

and by definition the thermal efficiency, or conversion ratio η_{Th} is given by

$$\eta_{Th} = \frac{\dot{W}}{\dot{W} + \dot{Q}_R} \qquad [5.2.2]$$

We now define the combustion efficiency η_c by the relationship

$$\eta_c = \frac{\dot{Q}_F - \dot{Q}_E}{\dot{Q}_F} \qquad [5.2.3]$$

which clearly is less than unity even if the combustion process releases all the available energy in the fuel. Using equations [5.2.1], [5.2.2] and [5.2.3] we find each energy transfer rate in terms of $\dot{Q}_F$ and get

$$\dot{Q}_E = (1 - \eta_c)\dot{Q}_F, \qquad \dot{Q}_R = \eta_c(1 - \eta_{Th})\dot{Q}_F \quad \text{and} \quad \dot{W} = \eta_{Th}\eta_c\dot{Q}_F$$

The rate at which the plant makes a profit $\dot{p}$ is given by

$$\dot{p} = \alpha\dot{W} + \beta\dot{Q}_R - \gamma\dot{Q}_F - \delta\dot{Q}_E - \varepsilon\dot{Q}_F$$
$$= \{\alpha\eta_{Th}\eta_c + \beta\eta_c(1 - \eta_{Th}) - \gamma - \delta(1 - \eta_c) - \varepsilon\}\dot{Q}_F \qquad [5.2.4]$$

The total operating costs of the plant comprise three costs, fuel, operation and amortisation, and finally the assumed penalty for producing an exhaust emission. Hence the profitability of running the plant is

$$P = \frac{\dot{p}}{(\gamma\dot{Q}_F + \delta\dot{Q}_E + \varepsilon\dot{Q}_F)}$$

or

$$P = \frac{\{\alpha\eta_{Th}\eta_c + \beta\eta_c(1 - \eta_{Th}) - \gamma - \delta(1 - \eta_c) - \varepsilon\}}{\{\gamma + \delta(1 - \eta_c) + \varepsilon\}} \qquad [5.2.5]$$

Equation [5.2.5] is the basic thermo-economic equation describing the plant operation. It takes into account environmental effects through the exhaust cost, and, if the heat rejection $\dot{Q}_R$ is a net liability, then this can be taken into account by inserting a negative value for β.

Example 5.2.1
A steam power plant similar to that of Fig. 5.2.1 is to be installed on a remote island in an Arctic region. The plant is to supply all the power and heating requirements for the island, and all heat rejection $\dot{Q}_R$ is to occur in houses using a district heating scheme. The plant is supplied and operated free under an aid agreement, and the only cost is that of fuel. Exhaust emission is regarded as a curiosity rather than a liability. Given that the combustion efficiency is 0·8 and that energy requirements dictate a ratio $\dot{W}/\dot{Q}_R = 0\cdot4$, find the profitability function for the plant and consider various arrangements for making energy charges.

Data

$$\frac{\dot{W}}{\dot{Q}_R} = 0\cdot4, \qquad \delta = 0, \qquad \varepsilon = 0, \qquad \eta_c = 0\cdot8$$

Analysis and calculation

$$\eta_{Th} = \frac{\dot{W}}{\dot{W} + \dot{Q}_R} = \frac{0\cdot4}{(0\cdot4 + 1\cdot0)}$$

$$= 0\cdot286$$

With the data inserted equation [5.2.5] thus becomes

$$P = \frac{1}{\gamma}\{(\alpha \times 0\cdot286 \times 0\cdot8) + (\beta \times 0\cdot8 \times 0\cdot714 - \gamma)\}$$

$$= 0\cdot229\left(\frac{\alpha}{\gamma}\right) + 0\cdot571\left(\frac{\beta}{\gamma}\right) - 1$$

Thus, for profitability,

$$\frac{0\cdot229\alpha + 0\cdot571\beta}{\gamma} > 1\cdot0$$

Examples of possible charging schemes and the consequent value of P are given in the following table.

$\left(\dfrac{\alpha}{\gamma}\right)$	$\left(\dfrac{\beta}{\gamma}\right)$	P
2·0	2·0	0·6
2·0	1·5	0·315
2·0	1·0	0·029
1·5	1·5	0·2
1·5	1·0	−0.08
5	0	0·145

A significant proportion of our energy resources, particularly in cold climates, is used for heating of offices, shops and homes. At first sight this is curious since the human body is itself a combustor of fuel in the form of food, and in principle, providing it has sufficient thermal insulation, there should be no requirement for the use of an external energy source. It is not necessary for the insulation material to be in contact with the surface of the body. Well-insulated walls of a large box will be sufficient to ensure satisfactory temperatures for a person inside it. Unfortunately, human beings need to breathe oxygen and after a while in a sealed room they would also find the odour rather objectional. On both counts ventilation is a necessity.

The ideal room for an energy-prudent human being is shown in Fig. 5.2.2. The room is perfectly insulated so that there is no heat transfer through the surface, and the mass flow rate of air $\dot{M}$ is just sufficient to stop odour or other problems. Inlet air is at temperature T_1, and the outlet is at what would be a comfortable room temperature, say 20 °C. The energy input to the room is $\dot{Q}$ to maintain steady conditions. Let us now calculate a suitable value for $\dot{Q}$ assuming a room volume of 60 m³, a complete change of air every hour, and an outside air temperature of 10 °C. An energy balance gives

$$\dot{Q} = \dot{M}c_P(T_2 - T_1)$$
$$= \frac{60 \times 1\cdot225 \times 1\cdot0(20 - 10)}{60 \times 60}$$
$$= 0\cdot204 \text{ kW}$$

The average human being at rest produces a heating effect of at least 0·15 kW, and consequently it can now be seen that for one person in the room, the external energy requirement is of the order of only 50 W and with two people the requirement is for *cooling* and *not* heating of the room.

When we consider extreme external temperatures of say −10 °C the energy balance then gives an energy transfer from the room of 0·612 kW and then, for one person, the basic heating requirement is

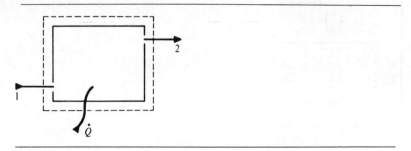

Fig. 5.2.2. The ideal thermodynamic room.

still only of the order of $0.5\,\text{kW}$ decreasing to $0.3\,\text{kW}$ for two persons.

It can be seen from these figures that energy consumption in the home and office, for the purpose of maintaining satisfactory temperatures, can be made quite small by careful design of insulation and ventilation.

5.3 Efficiency of conservation processes

In this chapter we have shown how a typical power plant (Fig. 5.2.1) can be investigated and the economics of its entire operation determined. In effect, the profitability given by equation [5.2.5] is a measure of the financial efficiency of the plant within the society and environment in which it exists. The analysis does not in any sense determine the question of whether – having regard to the requirement to conserve energy resources – it should exist in the first place. A detailed examination of the energy required to manufacture all the components in the plant, to construct the plant and maintain it, might prove to be greater than the total flow of energy through the plant over its useful service life. If this was shown to be the case, then it might appear better to build a simpler plant requiring less energy to build and accept more energy rejection from the system over its operating life.

This discussion indicates that a financial investigation of the operation of complex plants may well prove to be insufficient in order to protect energy resources, and that energy balances should include the energy required to construct, as well as to operate, the system.

Example 5.3.1
A small domestic lawn-mower is driven by a water-cooled engine in which the coolant is circulated round the cylinder head by an engine-driven pump. A well-intentioned 'do-it-yourself' magazine advises that the fuel consumption can be

improved if the pressure loss in the coolant system is reduced by polishing the internal passages. A keen energy
conservationist does this using a 500 W electric drill and the
job takes him 2 hr. If the rate of flow of cooling water is
0·1 kg/sec and the reduction in pressure drop achieved is
10 kN/m², determine whether the practice should be banned
on the basis of energy conservation.

Analysis and calculation
Energy used by drill in 2 hr is

$$2 \times 3600 \times 500 \text{ J} = 3 \cdot 6 \text{ MJ}$$

Assume power station conversion ratio of 0·3 and generation
and distribution losses of 10% gives a total energy requirement for drill operation of

$$\frac{3 \cdot 6 \times 1 \cdot 1}{0 \cdot 3} = 13 \cdot 2 \text{ MJ}$$

For the cooling systems we have a reduction in dissipation
of $\Delta \dot{\Phi}$ given by

$$\Delta \dot{\Phi} = \frac{\dot{M}}{\rho} \Delta p$$

where Δp is the reduction in pressure drop. Hence

$$\Delta \dot{\Phi} = \frac{10 \times 0 \cdot 1}{10^3} \text{ kJ/sec}$$

$$= 1 \text{ J/sec}$$

Assume petrol engine has an overall efficiency of say, 30%,
then the true energy-saving rate $\dot{E}$ is

$$\dot{E} = \frac{1 \cdot 0}{0 \cdot 3}$$

$$= 3 \cdot 33 \text{ J/sec}$$

If t is the time required to break even on the energy balance
we have

$$t = \frac{13 \cdot 2 \times 10^6}{3 \cdot 33}$$

$$= 3 \cdot 96 \times 10^6 \text{ sec}$$

$$= 1100 \text{ hr}$$

If we assume that cutting the lawn takes 1 hr and is done
say thirty times a year, then the number of years of oper-

ation before break-even is obtained is $1100/30 = 36$ years. Clearly the life of the machine is much less than this and the energy balance would never break even. The practice should therefore be banned.

5.4 Integrated energy systems

Most large-scale industrial processes require a combination of heat input and work input. In some cases, such as the production of bricks from clay, the dominant energy-transfer mode is heating. In contrast, the assembly of motor cars from their components is largely dependent on a work rather than a heat input. It might at first seem logical to consider whether a brick producer and a car manufacturer whose plants were sited close to each other, might economically be supplied with their differing energy requirements by some integrated power-plant system in which the work output was used by the car manufacturer and the heat rejection by the brick producer. Our studies of the Carnot theorem and power-plant cycles show that this is an unsatisfactory proposition in this case, because heat rejection occurs at *low* temperatures for reasonable conversion ratios and the brick manufacturer requires heat intake at *high* temperatures for a quality product.

This example illustrates the dilemma facing the designer of an integrated energy-supply system. High conversion ratios dictate low-temperature heat rejection, and high-temperature heat rejection necessarily carries the penalty of a low conversion ratio. As a consequence, integrated energy systems would appear to be worth consideration only when the process requiring a heat input can be effective with the energy input at reasonably low temperatures, well below the maximum temperatures found in thermal-power-plant cycles. This is generally true for thermal power plants where the energy supply comes from the combustion of a fuel, coal, oil or gas. The situation is rather different when the energy supply is obtained from a nuclear reactor. As discussed briefly in section 4.4, a nuclear reactor power plant is economic only when the reactor is used on a continuous basis at or near its maximum loading, and consequently it is generally found in base load generating stations. In some cases the station must be designed to carry much of the additional peak loading – for example if it is the main generating plant in an isolated community – and in these circumstances, it may be desirable to use the excess capacity for other purposes outside the peak period. Many chemical and allied processes are ideally suited to be run intermittently or at varying outputs so that they can adapt to the varying excess thermal capacity of a nuclear generating station.

The discussion so far has indicated two quite distinct areas where an integrated system may prove to be a viable proposition, and these are shown schematically in Fig. 5.4.1. The integrated system shown in (a) is appropriate for a combustion power plant combined with a process plant in which the demand varies little with time. In contrast Fig. 5.4.1(b) shows a possible scheme whereby a nuclear reactor can be run at near constant load, regardless of variations in generating requirements. This latter scheme requires a process plant where the output, and hence energy requirements, can be adjusted to suit the excess energy supply $(1 - \alpha \dot{Q}_S)$ available from the reactor. Obviously there are many possible variations between the two extreme cases shown in Fig. 5.4.1.

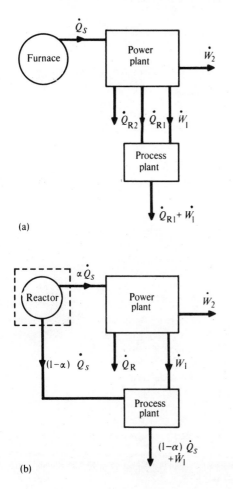

(a)

(b)

Fig. 5.4.1. Integrated energy systems.

Example 5.4.1
A coal-fired generating station and a process plant are to be built on adjoining sites and an integrated energy scheme is being considered. The figure below shows the separate and combined schemes being considered.

The process plant requires a heat supply at a temperature above that of the normal rejection temperature of the power plant, and in providing this the conversion ratio ($\dot{W}/\dot{Q}_3$) would be likely to reduce from 0·26 to 0·20. If $\dot{Q}_1 = 200$ MW and $\dot{Q}_3 = 70$ MW determine which scheme has the lower fuel consumption, assuming that boiler and combustion efficiencies are the same in the power station and process plant.

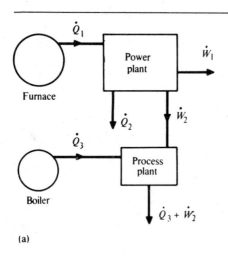

(a)

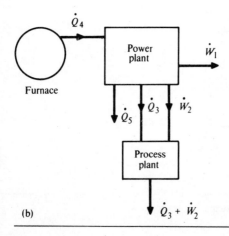

(b)

Data

As in Fig. Ex. 5.4.1 and with

$$\dot{Q}_1 = 200 \text{ MW}, \qquad \dot{Q}_3 = 70 \text{ MW}$$

$$\eta_a = 0 \cdot 26, \qquad \eta_b = 0 \cdot 20$$

Analysis

$$\dot{W}_1 + \dot{W}_2 = \dot{Q}_1 \eta_a$$

Therefore

$$\dot{Q}_4 = \frac{\dot{W}_1 + \dot{W}_2}{\eta_b} = \dot{Q}_1 \left(\frac{\eta_a}{\eta_b} \right)$$

Therefore energy supply for scheme (a), $\dot{Q}_a$, is given by

$$\dot{Q}_a = \dot{Q}_1 + \dot{Q}_3$$

for (b) the corresponding value is $\dot{Q}_b$ given by

$$\dot{Q}_b = \dot{Q}_4$$

$$= \dot{Q}_1 \left(\frac{\eta_a}{\eta_b} \right)$$

Calculation

$$\dot{Q}_a = (200 + 70) = \mathbf{270 \text{ MW}}$$

$$\dot{Q}_b = \frac{200 \times 0 \cdot 26}{0 \cdot 2} = \mathbf{260 \text{ MW}}$$

Thus scheme (b) has the lower fuel consumption, a reduction compared to (a) of approximately 4%.

Integrated energy systems can also be valuable on a much smaller scale than that considered in Example 5.4.1, particularly where the remoteness from conventional energy sources is a main problem. In such cases it can be worth while having a number of energy resources and combining them in a system suitable for the various energy requirements. Consider, for example, the problem of a remote communications station where the natural energy resources and energy requirements are as listed below:

Energy resources	Energy requirements
Good regular sunshine record	Electrical supply for electronics
Moderately windy on most days and	and lighting
nights	Power for water pump
Methane from bird droppings	Power for air-conditioning
	of building and electronics
	Heating for building during
	cold periods

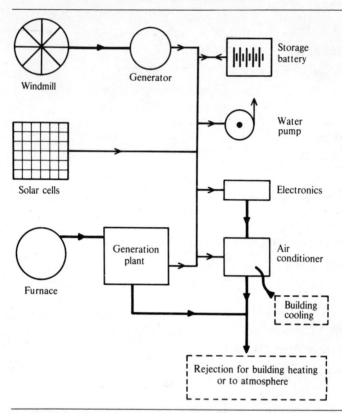

Fig. 5.4.2. Energy-supply scheme.

In Fig. 5.4.2 we see one possible integrated system which could provide the station requirements and which takes account of the lack of sunshine at night. It is clear that many variations in the arrangement of components in this system can be made; also, the relative amounts of energy flux in the energy-distribution network can be made within the overall limitations imposed by the energy resources and requirements of the station. In practice a number of alternative schemes would have to be analysed and compared on the basis of reliability, capital and running cost, ease of maintenance, etc. The analysis involved can be seen to be quite complex, even for the comparatively straightforward energy requirements of this example.

Example 5.4.2
A remote farm is equipped with a diesel-powered generation set of 5 kW electrical output and a windmill with a self-contained electrical generator providing an output of 2 kW.

The two sets of generation equipment have compatible electrical characteristics and may be used together or separately, and at reduced power outputs, with little loss in overall efficiency. The total capital, maintenance and running costs of each system are 0·9p for the windmill and 2·2p for the diesel set – in each case for 1 kWhr of electrical power output. During a typical 24 hr day the electrical requirements of the farm are as follows:

Load (kW)	Hours
6	2·5
4	1·5
3	12.0
2	8.0

Assuming the windmill can be operated at any time during the period, find the minimum energy cost for the 24 hr period and the saving achieved by having the windmill generator.

Analysis and calculation
The windmill can supply a base load of 2 kW during the full 24 hr period. Therefore,

$$\text{Windmill output} = 2 \times 24 \text{ kWhr}$$

$$\text{Windmill output cost} = 2 \times 24 \times 0·9$$
$$= 43.2\text{p}$$

$$\text{Generator output} = \{(4 \times 2·5) + (2 \times 1·5) + (1 \times 12)\}$$
$$= 25 \text{ kWhr}$$

$$\text{Generator output cost} = 25 \times 2·2$$
$$= 55\text{p}$$

Hence, minimum cost for 24 hr period $= 43·2 + 55$
$$= 98·2\text{p}$$

With generator supplying all electrical power its output = 73 kWhr therefore:

Cost of generator supplying all the power $= 73 \times 2·2$
$$= 160·6\text{p}$$

Hence, saving = **62·4p/day**

In percentage terms the saving $= \dfrac{62·4}{160·6} \times 100$

$$= \textbf{38·9\%.}$$

Example 5.4.3

The farmer of Example 5.4.2 is offered a second identical windmill at a total installed cost of £1200 (£1 = 100p). He borrows the money at a true interest rate of 9·5%, taking tax advantages into account. Maintenance charges for the new set amount to £60 per annum.

Assuming his daily energy requirements remain the same, would it be to his advantage to have the additional windmill installed?

Analysis and calculation

$$\text{Total cost of new set over one year} = £\left(60 + \frac{9 \cdot 5 \times 1200}{100}\right)$$

$$= £174$$

$$\text{Additional cost per day} = \frac{174 \times 100}{365}$$

$$= 47 \cdot 7\text{p}$$

Of the 24 kWhr previously supplied by the diesel generator, the amount that can now be supplied by the second windmill is given by

$$W = \{(2 \times 2 \cdot 5) + (2 \times 1 \cdot 5) + (1 \times 12)\}$$

$$= 20 \text{ kWhr}$$

Hence generator supplies $= (25 - 20)$

$$= 5 \text{ kWhr}$$

Cost of generator output $= 5 \times 2 \cdot 2$

$$= 11\text{p}$$

Total cost of supplying power requirement by new system:

$$(43 \cdot 2 + 47 \cdot 7 + 11) = 101 \cdot 9\text{p}$$

The new system will be more expensive by

$$(101 \cdot 9 - 98 \cdot 2) = \mathbf{3 \cdot 7\,p/day}$$

The cost balance is consequently in favour of the old system and he would be advised not to purchase the additional windmill for generation of electricity.

The previous example illustrates clearly the point that what may seem superficially to be a good idea, may, when analysed carefully, be seen to be a disadvantage. A full economic assessment is essential to this type of problem, and it is quite impossible to conclude anything unless this is done.

5.5 Energy storage

The lead–acid battery is a simple example of an energy-storage device giving the facility for charge and discharge at different times and, because of its portability, at different places if this is required. Unfortunately, this type of battery is in most cases unsuitable for use as the main power supply in road vehicles because of its bulk and weight, and for large-scale energy storage it proves to be far too costly. Other types of energy-storage devices exist which do not rely on an electrochemical process, and these include compressed-air bottles, flywheels and high-altitude water-storage reservoirs. All of these, and the lead–acid battery, have the common feature that the output process is not a thermal energy transfer and thus a power output can be obtained quite independently of any limitations imposed by the Carnot theorem. On the other hand, if energy was to be stored by heating material to a high temperature and later recovering the energy in a cooling process, then the engine required to obtain a power output would necessarily reject a significant proportion of the energy storage.

It is interesting now to compare the relative volume and mass of various energy-storage devices on the basis that the energy storage required in each case is 1 MJ. The figures given in Table 5.5.1 have been rounded off to make comparison easier.

It is clear from this table that large-scale storage for power production is only viable at present by method (c) because the enormous volume of storage required makes a water reservoir the only practical storage tank. If, for example, it was required to store the output of a 300 MW power station over a period of 1 hr, then the volume of water required is of the order of $3 \times 10^5 \, m^3$ which is not large. The corresponding storage volume for method (b) is $2 \times 10^4 \, m^3$, which is becoming a sizeable pressure vessel. It may be possible in the future to build large-scale pressure vessels at depth in the earth's rock, which could then make this method attractive. Table 5.5.1 also shows that the electric battery is most suitable for storage of energy when a small power output is required, as in starting a motor car. For large vehicles compressed air is often used for starting purposes, and Table 5.5.1 shows that this will involve a similar volume of storage space with a saving in weight compared to

Table 5.5.1

Storage method	Mass (kg)	Volume (m^3)
(a) Lead–acid battery	2×10	2×10^{-2}
(b) Compressed air at 15 MN/m^2	8×10^{-1}	2×10^{-2}
(c) Water at an altitude of 300 m	3×10^2	3×10^{-1}
(d) Aluminium at a temperature of 600 K	4	$1 \cdot 5 \times 10^{-3}$
(e) Steam at atmospheric pressure	4×10^{-1}	7×10^{-1}

the electric battery. Thermal energy storage by methods (d) or (e) is, in comparison, costly on either mass or volume, and power production subject to limitations on the achievable conversion ratio. Where only a thermal output is required – as in home or office heating – method (d) is preferable to (e), providing the costs of storage equipment are offset by some advantage obtained from the method of costing the energy input at different periods during the day.

Table 5.5.1 enables rough comparisons to be made between alternative storage methods for particular applications, but can only serve as a guide. Strictly, the comparison can only be made by taking all the economic circumstances into account and including all the dissipation effects in the storage and recovery processes.

Example 5.5.1
A nuclear power station is to be run on a continuous basis supplying 800 MW throughout the full day. During an 8 hr period in the night the output power is 110 MW greater than the demand, and it is proposed to use this power output to pump water from a large lake to a smaller lake 380 m higher up. Later, during the peak demand period, the water will return to the lower lake through a turbine generator set and thereby boost the generation capacity. If the overall efficiency of both the storage and recovery process is 85%, estimate a minimum capacity for the upper lake and the energy dissipated in the complete process.

Data

$h = 380 \text{ m}$

$W = 110 \times 8 \text{ MWhr}$

$\eta_S = 0.85$

$\eta_R = 0.85$

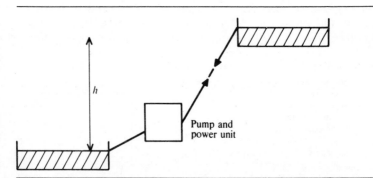

Analysis

Let volume of water required = V, then the work required to raise the water W' is given by

$$W' = \rho V g h$$

where g is the local gravitational acceleration. The work input W required to do this is therefore

$$W = \frac{W'}{\eta_S} = \frac{\rho V g h}{\eta_S}$$

and hence

$$V = \frac{W \times \eta_S}{\rho g h}$$

The electrical energy E that can be recovered is therefore

$$E = \eta_R W' = \eta_R \eta_S W$$

and the dissipation Φ is consequently given by

$$\Phi = W - E = W(1 - \eta_R \eta_S)$$

Calculation

$$W = 110 \times 8 \times 3600 \text{ MJ}$$
$$= 3 \cdot 168 \times 10^{12} \text{ J}$$

$$V = \frac{3 \cdot 168 \times 10^{12} \times 0 \cdot 85}{1000 \times 9 \cdot 81 \times 380}$$
$$= 722 \times 10^3 \text{ m}$$

$$\Phi = 3 \cdot 168(1 - 0 \cdot 85^2) \times 10^{12} \text{ J}$$
$$= 879 \times 10^3 \text{ MJ}$$

Example 5.5.2

A cylindrical iron flywheel is 1 m in diameter and rotates at 1000 r.p.m. Compare this flywheel with the lead–acid battery as a means of providing a power output to drive a milk delivery vehicle over short distances. Assume the density of iron is $7 \cdot 88 \times 10^3$ kg/m^3.

Data

$$d = 1 \text{ m}$$

$$\omega = \frac{1000 \times 2\pi}{60} \text{ rad/sec}$$

Analysis

$$(k.e) = \int_{r=0}^{r=R} 2\pi r \, dr \rho l (r\omega)^2$$

$$= \frac{\pi \rho l \omega^2 d^4}{32}$$

Volume of flywheel $= \dfrac{\pi l d}{4}$

Mass of flywheel $= \dfrac{\pi \rho l d^2}{4}$

Calculation

$$(k.e) = \frac{\pi \times 7 \cdot 88 \times 10^3}{64} \left(\frac{1000 \times 2\pi}{60}\right)^2 \times 1^4 \times l$$

$$= 2 \cdot 12 \times l \times 10^6 \, \text{J}$$
$$= 2 \cdot 12 \times l \, \text{MJ}$$

Hence for 1 MJ energy storage

$$l = \frac{1}{2 \cdot 12} = 0 \cdot 472 \, \text{m}$$

Therefore

$$\text{volume} = \frac{\pi \times 0 \cdot 472 \times 1^2}{4}$$

$$= 3 \cdot 7 \times 10^{-1} \, \text{m}^3$$

$$\text{Mass} = 3 \cdot 7 \times 10^{-1} \times 7 \cdot 88 \times 10^3$$
$$= 2 \cdot 92 \times 10^3 \, \text{kg}$$

Comparison with the data for the lead–acid battery in Table 5.5.1 shows that, for the same energy storage, the flywheel is far heavier and takes up more volume than the battery. Unless the flywheel and associated equipment prove to be very much cheaper to install and maintain – which is unlikely – the battery will prove to be the preferred energy-storage device.

5.6 Problems

1. A small factory produces a wool material by weaving, and the steam power plant it has previously used breaks down and cannot be repaired. The decision is taken to replace the plant by a water turbine in the river close to the factory. The costs

involved include a charge levied by the river authority proportional to the amount of water consumed, capital and running costs of the turbine unit, and a likely tax penalty from the effect of the annoying noise of the turbine upon the local community. Assess the effect of these factors upon the plant economics.

2. The manufacture of concrete involves mixing mechanically stones, sand, cement and water. The mixing process takes a significant time and the energy used in the process is dissipated and does not assist in the chemical reaction. Examine alternative methods of producing concrete and try to make a rough assessment of energy savings that might be made.

3. Saturated steam at $2000 \, kN/m^2$ pressure is available in practically unlimited quantities. It is proposed to use the steam for power generation, using a river 100 m below the volcanic site for cooling purposes. The river has a temperature of 288 K, and a maximum flow of $2 \times 10^6 \, kg/hr$. Find the maximum possible power output of the station, assuming that the downstream temperature of the river must not be allowed to exceed 293 K.

   An alternative proposal is made that the river water should be used to cool the steam at source, the condensate then being piped down to a hydraulic turbine near the river. Would this be a better scheme?

Answers to problems

Chapter 1

2 40 mm
4 785 kN/m^2; 39·6 m/sec
7 928 kW

Chapter 2

3 1037 MJ
5 No

Chapter 3

2 1·93 cm^3; 0·680 J
3 54 W
7 119 kN/m^2; 12·1 kN/m^2; 315 K; 216·7 K
10 Hint – check change in entropy

Chapter 4

1 1·399 MW; 3·073 MW
3 0·101
6 3·82 W

In a number of cases slightly different answers can be obtained depending on the assumptions made.

Index